Minding Dogs

ANIMAL VOICES
ANIMAL WORLDS

Robert W. Mitchell, series editor

SERIES ADVISORY BOARD

Jonathan Balcombe

Margo DeMello

Francine L. Dolins

Hal Herzog

Dale Jamieson

Claire Molloy

Paul Waldau

Sara Waller

Minding Dogs

HUMANS, CANINE COMPANIONS,
AND A NEW PHILOSOPHY OF
COGNITIVE SCIENCE

Michele Merritt

The University of Georgia Press

ATHENS

© 2021 by the University of Georgia Press
Athens, Georgia 30602
www.ugapress.org
All rights reserved
Set in 10.25/13.5 Kepler Std Regular by
Kaelin Chappell Broaddus

Most University of Georgia Press titles are
available from popular e-book vendors.

Printed digitally

Library of Congress Cataloging-in-Publication Data

Names: Merritt, Michele, 1980– author.
Title: Minding dogs : humans, canine companions,
   and a new philosophy of cognitive science / Michele Merritt.
Description: Athens : The University of Georgia Press, [2021] |
   Series: Animal voices : animal worlds |
   Includes bibliographical references and index.
Identifiers: LCCN 2020041742 | ISBN 9780820359533 (hardback)
   | ISBN 9780820359557 (paperback) | ISBN 9780820359540 (ebook)
Subjects: LCSH: Dogs—Behavior. | Cognition in animals. |
   Dogs—Psychology.
Classification: LCC SF433 .M47 2021 | DDC 636.7—dc23
LC record available at https://lccn.loc.gov/2020041742

CONTENTS

INTRODUCTION 1

CHAPTER 1 Canine Minds: Historical Precedents and Current Curiosity 10

CHAPTER 2 Thinking-with Dogs and Dismantling Standard Cognitive Science 33

CHAPTER 3 Canine Mindreading and Interspecies Social Cognition 66

CHAPTER 4 Thinking-in-Playing: Social Cognition beyond Mindreading 97

CHAPTER 5 Dynamic Duos: Making-with, Thinking-with, and Enacting Interspecies Collaborations 125

POSTSCRIPT Six-Million-Dollar Dogs and the Future of Companion Species Studies 153

NOTES 161

WORKS CITED 167

INDEX 189

Minding Dogs

INTRODUCTION

In 1993 Don Smith, while shooting film for CBS News in St. Louis during the "Great Flood" of the Mississippi and its tributaries, saw a puppy clinging to life in the rushing waters. Smith and his crew were at a safe elevation but were close enough to capture the horrific scene as it unfolded. In the video the puppy, who appeared to be no more than eight weeks old, was trying to climb to higher ground but constantly getting washed away. The pup finally climbed onto a pallet and caught a brief respite, but the waters quickly rose again. Even cars and sheds were being carried away.[1]

At one point Smith set his camera down and waded into the water to try to save the little dog. But a massive tree floated by, just missing him, and he had to retreat. When the water surged over the pallet, Smith saw that the dog was now heading toward a spot where he could make one last attempt at a rescue. Lunging along a flooded roadway, Smith almost miraculously reached the small dog and brought it to safety.

Watching this video, or any of the countless others like it, arouses all sorts of emotions, from fear and anxiety to relief and tears of joy. The internet has become a repository for such animal videos, and in turn, commentary on these vignettes provides a philosopher with countless examples of folk psychology gone wild. People find it irresistible to explain animal behavior in humanlike mentalistic terms—a dog was *relieved* to see a human, a crow *befriended* a little girl by giving her gifts, or the elephant *grieved* the loss of its mother.[2] Folk psychology, which can be thought of most simply as a common-sense approach to understanding cognition, is not necessarily unscientific. As Allen and Bekoff argue, "the generalizations and theoretical terms of folk psychology may be suitably refined and incorporated into a fully scientific theory of mind and behavior applicable to both humans and nonhumans" (1997, 66). Nevertheless, the anthropomorphism rampant on the internet is often unwarranted, or, at the very least, has not been convincingly supported by scientific evidence. As any respectable scientist or philosopher will attest, one compelling YouTube video does not suffice to prove that an animal possesses this or that cognitive capacity.

Don Smith's experience likewise cannot be used as scientific evidence that the puppy felt relief on seeing the man wading to the rescue or inherently trusted him. Yet I begin with this story because it highlights something people intuitively know about dogs: they are highly dependent on humans, and sometimes their lives completely depend on us. Furthermore, the video of the rescue of "Rescue" (the puppy was adopted by the helicopter pilot and named accordingly) offers some moments worth pausing over, considering all we have learned about dogs over the last decade or so. First, Smith's inability to just let the dog drown was so overwhelming that he put his own life at risk. This speaks to the love many humans feel for dogs—a love that rivals our love for human children (Levin et al. 2017). Second, the puppy was clearly growing exhausted and at one point seemed to be giving up, but when he could hear Smith's voice calling and drawing closer, a spark of resilience was ignited and Rescue mustered enough energy to fight the current and swim toward Smith. It is hard not to think that the puppy just knew he needed to trust Smith and get to him as best he could.

Humans and dogs indeed have a unique bond, unlike so many other interspecies relationships, which is perhaps unsurprising given that we have been closely intermingling for well over fifteen thousand years. Perhaps this is at least part of the reason Rescue seemed to "just know" to head toward the human voice he heard above the rushing waters. This might also explain why Smith found it impossible not to try to save him. Like many readers of this book, I have always believed that the dogs with whom I share my life love me, and I believe that they understand quite a bit about me and what I am thinking and that they know a great deal about humans in general. Yet it wasn't until I read about Chaser, the border collie who had been dubbed the "smartest dog in the world" (Cooper 2015), that I realized that scientists were starting to demonstrate what so many of us have believed to be true of our canine companions. At the time, I was working on my dissertation, a treatise that had nothing to do with dogs, or so I thought. It was a defense of a theory in the philosophy of cognitive science generally referred to as externalism—the idea that the mind, or cognition, is not something occurring solely inside the head but is rather a process that is constituted by organism-environment couplings. Most notably, my dissertation focused on the ways in which human-technology couplings could constitute thinking, such as the way my smartphone is arguably an extension of my fleshly body, a machine that thinks *with me*. When I read about Chaser, and subsequently all the other fascinating studies coming out of canine cognition labs, I began to wonder: *To what extent are dogs extensions of us?*

That question was the genesis of this book project. It has been many years in the works, as I slowly pieced together the two seemingly disparate areas of my research: on the one hand what I have defended as *radical philosophy of cognitive science*, which sees thinking as an extended, enacted, and ecological process, and on the other the growing work in cognitive ethology that overwhelmingly suggests that not only do dogs think in sophisticated ways heretofore overlooked by researchers, but they do so, at least in part, as a result of hanging around with us for so many years. Much as the radicals in philosophy of cognitive science claim that some forms of cognition can be properly understood only by including the active environmental engagements as *part of those cognitive processes*, so too, I argue, do dogs and humans form collaborative pairs in which unique forms of cognitive processes emerge. These forms of thinking cannot be fully understood by utilizing a subjective and internal model of cognition. Dogs and humans are also conjoined emotionally and socially—dogs are masterful decoders of human affect and intent. And we *play together*. Agility and flyball, even choreographed dog-human dance competitions, are some of the more formalized games humans and dogs play, but even if your dog is not a well-trained agility champion, you have likely engaged in all manner of spontaneous play together. Play, it turns out, is perhaps the point at which the two species *Homo sapiens* and *Canis familiaris* know each other best, and where genuinely unique forms of thinking emerge and are sustained.

This book is an examination of the dyadic exchanges between humans and dogs and what those interactions have afforded both species in terms of how cognition is constituted. As a philosopher, I have an overarching set of arguments I am defending. Primarily I aim to continue defending the radicals in philosophy of cognitive science, in particular the enactivist account of cognition, and a close inspection of human-dog dyads is a novel and compelling way to do just that. Philosophers have written about nonhuman animal minds, but taking seriously what the interspecies relationship between humans and dogs might mean for the philosophy of mind and cognitive science has, until now, been ignored. This project is therefore an attempt to once again rethink the boundaries of cognition—what it means to think and where that thinking occurs—but also to rethink the boundaries of ourselves, and what is possible when two species collaborate for so many years. As I am not a trained scientist, I do not attempt to rethink how canine cognition studies are undertaken per se, although I hope this book is appreciated not just by philosophers but by those in related disciplines as I provide an alternative framework from which to approach the study of dogs. The implications of what I argue extend to other domains related to dogs, including issues pertaining to animal welfare, contemporary ethical problems surrounding animals, and responsible pet guardianship.[3]

In the first chapter I lay some groundwork for the project I am undertaking and clarify my aims and methodology. A brief overview of the historical underpinnings of cognitive ethology raises a question: To what extent can we hope to genuinely study the cognitive capacities of dogs in their "natural" habitats? I examine the concept of an *umwelt* or "lifeworld," as many philosophers have dubbed it, and how this concept is employed both in ethology and in philosophy. Upon carefully considering the lifeworld of dogs,[4] the supposedly firm dichotomy between animals in their "natural" habitats and domesticated animals begins to crumble because, for many dogs, what is "natural" *is* domesticated life and, moreover, human interaction. So the idea that human interference tarnishes any credibility we might give to ethological studies of dogs is an argument I reject. To be sure, it is wise to exercise caution in accepting these findings as definitive demonstrations of any functional similarity between the ways humans and dogs cognize. Just because a dog follows a human's pointing gesture to find food, even when that gesture is misdirecting the dog, it would be overly zealous to assume from this one finding that dogs have the capacity to understand others' mental states, or that they greatly trust their human companions to tell the truth. That a dog can form thoughts about the mental states of a human (or any other being, for that matter) or that a dog has concepts of "truth" and "trust"—these are capacities that dogs *might* turn out to possess, but we would need more evidence than a few dogs following pointing gestures to prove it. To this end, I utilize the idea of *critical anthropomorphism* (see Burghardt 1991) to guide our thinking about the findings many canine researchers have reported in the last decade. As I am coming to this project from a philosophy background, I am accustomed to skepticism regarding other minds, but I do not think it is impossible to form well-reasoned arguments about the nature of what dogs are thinking and feeling. Just as we must avoid egregious anthropomorphism, we must also resist uncritical anthropocentrism because it is quite probable that other animals think in ways that look nothing like how we do it. Proceeding with the realization that it is impossible not to attribute *some* humanlike traits to nonhumans, but that when we do, we must critically examine all the factors influencing our attribution and rule out competing interpretations and confounding variables before settling on an explanation, will be the central methodology of the book.

Chapter 2 delves right into one of the core arguments I am defending in this project: that dogs and humans think *together* in ways that traditional philosophy and cognitive science—which has largely adopted a cognitivist framework—are unequipped to explain. One of the corollary arguments I defend in this chapter is that "cognition" ought to include affect, despite what many philosophers over the years have argued. Thus I spend some time explicating the emotions and how they are integral to thought, a position that is overwhelmingly defended in some of the nontraditional strands in philosophy of cognitive science. Then I survey recent findings pertaining to canine cognition that show not only that dogs are surprising us with what they are capable of but that these skills are overwhelmingly *social* and mimic in many important ways the development of human infants as they learn to engage with others. Over the years, dogs have learned that a valuable partnership obtains with humans, and they have used this partnership to proliferate, prolong their lives, and acquire all sorts of useful skills. But this relationship is mutually beneficial, as I discuss, because humans are deriving physiological and psychological benefits including, I argue, *cognitive scaffolding* from their relationships with dogs. The idea of scaffolding is important in this chapter, as it frames a lot of the discussion surrounding what I call the "radicals" among philosophers of cognitive science. Specifically, the claim that the environment we reside in—which includes the social world—shapes, transforms, and is a part of the thought processes and affective transactions in which we engage, is integral to my argument about how best to explain the interactions humans and dogs have had over many thousands of years. There are several varieties of nonstandard or "radical" philosophies of cognitive science out there, but I think the *enactivist* approach is best suited to explaining how individuals from two distinct species can *think-with* and co-constitute cognitive processes—and, further, how the species themselves might come to change slowly over time from these sustained interactions. Toward the end of the chapter, therefore, I bring together all the research emerging from canine ethology with the enactivist approach to cognition and argue (1) that we can best understand how cognition emerges and is sustained through interspecies interactions by allowing that dynamic exchanges with features of the environment are genuine parts of cognition, and (2) that the case for an enactivist account of cognition is actually bolstered by considering interspecies cognition, which I refer to as *coactive cognition*, thereby dismantling the typical cognitivist framework employed by so many philosophers.

The argument in chapter 2—that dogs and humans coactively think and affectively co-attune themselves to one another—might lead one to wonder if dogs possess some of the same mindreading skills as humans. If a dog can use my gestures to figure out where food is located, or sense when I am sad, or detect subtle changes in my tone of voice, do these sorts of findings taken together indicate that dogs are able to form thoughts about the mental states of others, especially humans? Chapter 3 indeed broaches this subject, but it begins with the critically anthropomorphic caution introduced in chapter 1. The picture sketched up to this point is overwhelmingly in favor of letting cognition "go to the dogs," defending the idea that they think with us in sophisticated ways and have affectively co-attuned themselves so well with humans that it could reasonably be argued that at least some dogs are better at reading humans than humans themselves are.[5] However, taking these findings too far and suggesting that dogs have robust "theories of mind" would be met with warranted skepticism. Thus, in chapter 3, I first sketch the great mindreading debate in philosophy and cognitive science, and argue preliminarily that two of its biggest contenders—Theory Theory and Simulation Theory—fall short in capturing the full array of what we ought to think of as "social cognition" skills. Focusing instead on the way interactions are fundamental to social cognition—or what some refer to as *primary intersubjectivity*—provides a more comprehensive picture of all the ways humans think about other minds. I then turn to nonhuman animals and ask if mindreading is even empirically tractable (that is, a phenomenon we can reasonably observe, measure, and understand scientifically) in species other than humans. While I ultimately think the answer is yes, much work will need to be done to convince the naysayers. Part of what makes studying nonhuman animal mindreading so difficult, I argue, is not so much uncritical anthropomorphism as unwarranted anthropocentrism. Rather than insisting that all mindreading skills must conform to a human-specific model, I urge thinking of mindreading as occurring on a spectrum, including skills that ought to count as "rudimentary" or primary forms of social cognition. Much of the research we have seen on dogs to this point supports putting their social cognition skills somewhere on this spectrum. And it turns out that when we focus on primary intersubjectivity, rather than relying solely on theory-formation or simulation, the spectrum theory of social cognition skills is further supported, as it helps us understand how dogs and humans engage in interspecies mindreading without needing to appeal to an overly anthropocentric model of the skill.

I end chapter 3 with a suggestion: Most of the studies cited in support of canine mindreading focus on the social referencing capacities dogs demonstrate—in other words, their ability to utilize social cues from others to comprehend or respond to a situation. With dogs, these capacities are demonstrated primarily in laboratory settings, or those designed by researchers like Brian Hare who aim to study dogs from the comfort of their own homes.[6] While these findings are indeed indicative of some important social skills that dogs possess, there are other sorts of interactions where one can expect to find humans and dogs engaged in social cognition together. Thus I suggest we look in chapter 4 at *interspecies play* and the way this sort of dynamic exchange affords both dogs and humans a plethora of opportunities to engage in what I term *fundamental social cognition*. As I note at the beginning of the chapter, play is taken seriously by developmental psychologists, for it is almost unanimously agreed to stimulate cognitive growth and aid in the acquisition of all manner of skills. Chapter 4 therefore presents an in-depth look at the ways humans and dogs play together. I first examine the ways psychologists and philosophers have historically categorized play and then how ethologists catalogue nonhuman animals' play. The bulk of this chapter is, however, devoted to work by Robert Mitchell and his colleagues, who have provided some of the most compelling evidence from their case studies of spontaneous play between humans and dogs of the sort of fundamental social cognition I argued for in chapter 3. Through an examination of the way humans and dogs cocreate "play projects" and routines, I think there is a strong case for viewing interspecies play as a mode of enacted cognition. In fact, human-dog play is just the sort of example that many enactivists, myself included, have argued really highlights the dynamic and interactive nature of certain forms of thinking.

Interspecies play constitutes not only a mode of thinking-with but also, I argue, a mode of *making-with*, or what Donna Haraway has referred to as *sympoiesis*. Chapter 4 ends with the suggestion that the tightly coupled human-dog dyads that occasionally form during play offer yet another instance of how enactivism is better suited to explicating this coactive cognition, because the thinking-with (or, better, the playing-with) that takes place cannot be readily understood simply by examining the inner workings of either the human's or the dog's brain. In chapter 5, I utilize the *siphonophore* as symbolic of this tight intermingling, and I also borrow from biology to introduce a relatively new concept, the *holobiont*, to further illustrate Haraway's notion of "making-with." Siphonophores are multiorganism creatures that have formed symbiotic relationships so interdependent that they are often mistaken for one animal. A holobiont, according to Bordenstein and Theis (2015), can be thought of as a network of biomolecular structures that includes a host and all its associated microbes. The analogy is of course not perfect, and I am not suggesting we think of humans and dogs at play as a new hybrid species or anything so drastic. Instead, the comparison between human-dog dyads and siphonophores/holobionts is meant to draw our attention to the ways in which multiple species can and often do come together in collaborative groupings to think and create, and that this coupling results in modes of cognition that are not possible for any of the individuals on their own. In particular, when two or more organisms collaborate to create something—as in spontaneous play between humans and dogs—the resulting "game" is at once constrained by some of the implicit rules of interactions generally but also freely formed through a relatively unpredictable process.

Chapter 5 therefore seeks to reconcile the spontaneity and unpredictable nature of play and creativity that marks so much of human-dog coupling with the enactivist notion of *autopoiesis*. Since most of my argument rests on the contention that the enactivist account of cognition is best suited to explicating interspecies cognition, it is imperative that I address the autopoietic qualities of thinking that so many enactivists argue are crucial to understanding the theory. Autopoiesis—the capacity of an organism to self-organize and self-preserve, all while continually adapting and changing—seems to be at odds with the account of thinking to which my argument has lent itself up to this point, because spontaneous play and improvisational thinking are seemingly unpredictable and not subject to any boundary constraints. Nevertheless, a case can be made that even in spontaneous play, there are constraining features at work. Furthermore, a closer look at prediction and how it works in cognition reveals that for a system to be predictable, it need not be entirely regulated or patterned. If, given certain "priors" or background conditions, we can expect a set of behaviors or actions with relatively minimal error, we can safely call the system predictable. To explicate this idea more fully, I turn to Andy Clark's 2016 work on *predictive processing* and argue that, although this view of cognition seems to resemble a much more traditional, internalist, and cognitivist approach, at its core it retains the flavor of externalism Clark has become known for defending. This recent work is actually, I argue, much more in line with enactivism than his previous work, especially if we interpret his model of predictive processing along the lines of what Shaun Gallagher and Micah Allen (2016) call *predictive engagement*. That is to say, prediction, like other forms of thinking, is not something happening behind the confines of the skin and skull but is rather a mode of engaging the world, testing it out to see what it affords, and tweaking action in response to those exchanges. Gallagher and Allen attempt to show that enactivism need not be at odds with predictability, and I not only endorse this line of argumentation, but I extend it to my discussion of interspecies cognition and of play specifically. I further claim that the account of sympoiesis I have given regarding human-dog playful pairs is also not at odds with the enactivist story. In fact, enactivism is made stronger by considering that in some couplings, the resulting *thought-in-action* is precisely the sort of sympoietic *making-with* that we see so compellingly on display when humans and dogs play and think together.

CHAPTER 1

Canine Minds

HISTORICAL PRECEDENTS
AND CURRENT CURIOSITY

Who Let the Dogs In?

This book has two central goals. One is to examine the way dogs think alongside humans and how this mutually beneficial relationship has shaped both canine and human cognition. The other is to consider what this collaborative cognition might imply for how philosophers study cognition, both human and nonhuman, more generally. In order to successfully and comprehensively examine these issues, it is worth first setting the stage for understanding why such a project is relevant in the first place and how it came to be so. Dogs have not historically been the darlings of behavioral science. The great apes and dolphins have dominated the scene when it comes to conducting comparative cognitive science with humans, although that landscape has changed drastically in the last twenty years. Octopi, crows, and even fish now warrant serious consideration among scientists and philosophers, as more and more research indicates these animals are intelligent and perhaps even capable of moral reasoning.[1] So too have dogs begun to capture more attention, and I intend to discuss at length many of the findings that have prompted folks to change their views on canine intelligence. Despite the growing interest in the sciences, dogs remain mostly shunned by philosophers, aside from the occasional Facebook group devoted to celebrating dogs who live with philosophers,[2] or a very few papers that seek to incorporate canine science into philosophical inquiry (e.g., Merritt 2015a). It is my contention that this oversight is unfortunate, as there is a vast wealth of philosophical insight to be gleaned from a careful examination of the science of canine intelligence and the co-activities of humans and dogs.

Before beginning a philosophical inquiry into the nature of this important interspecies relationship and the failure of many traditional views of cognition within philosophy to properly account for it, it is important to look at the historical impetus behind the relatively nascent interest in canine minds. The surge in attention to dogs in the last couple decades is itself an intriguing story, one rooted in deeply held assumptions about what counts as "intelligence," what is the proper way to study animals generally, and what exactly is meant when scientists argue we ought to observe animals in their *natural* habitats. To develop an appreciation of how dogs have wriggled their way into the scientific spotlight lately, let us look at why they have heretofore been excluded. A good place to begin is right where an important shift in thinking about animal intelligence began to take hold in most scientific communities. That shift resulted in what is now known as cognitive ethology, a field of inquiry that is simultaneously in its infancy and already the subject of quite a few long-standing and seemingly intractable debates.

A Brief History of Cognitive Ethology

Don Griffin is the undisputed originator of the term *cognitive ethology*, and his 1984 book *Animal Thinking* arguably ushered in the field as it is now conceived. Having specialized in bat echolocation for the first half of his career, Griffin altered the direction of his research program when he became concerned that strict behaviorism was not a sufficient means of grasping the complexity and richness of nonhuman animal lives. What was missing, he argued in 1976 in *The Question of Animal Awareness*, was an account of the mental lives of animals, including their conscious states and processes. That book was a charge to the scientific world to answer some very difficult but important questions, questions that classical ethology had not attempted to tackle.

Classical ethology, albeit in Griffin's view falling short of being a comprehensive account of animal cognition, is nonetheless a crucial point on the trajectory toward Griffin's ideal. The pioneering work done by Nikolaas Tinbergen (1951), together with Konrad Lorenz and Karl von Frisch, brought modern ethology to the fore,[3] and animal science began to seriously question the validity of experiments conducted with animals in laboratory settings. The behaviorism of B. F. Skinner (1953) had largely dominated the study of animals for some time, but Tinbergen saw the results of those investigations as likely specious, given the artificial environments and stimuli of the experimental setup. Ethology, he argued, would instead seek to study animals contextually, in their natural habitats, and with minimal to no human interference. Tinbergen (1963) proposed four questions to be answered if we wish to fully comprehend animal behavior. First, what is the mechanism involved? That is, what is causing the behavior and how is it constructed, from the organismic to the molecular level? Second, what adaptive value does the behavior have? In other words, how does what the animal is doing contribute to its survival and reproductive success? Next, what is the ontogeny of the behavior? How does it emerge and how is it sustained over the life of the animal? Last, what is the phylogeny of the behavior? How has it evolved in the species generally? These questions, Tinbergen argued, cannot be adequately addressed by forcing an animal out of its natural environment and into a lab.

Of course, the ethological revolution in animal behavior owes a great deal to Darwin (1859, 1871). Tinbergen's four questions—particularly the adaptive and phylogenetic questions—are in many ways a tribute to Darwinian evolutionary biology. Thinking about Darwin's research can, in a roundabout way, help bring the other piece of Griffin's argument into focus a bit more. Cognitive ethology is the coupling of ethology with cognitive science, which until recently has been much more aligned with classical behaviorism in that many studies carried out in cognitive science have taken place in laboratory settings. However, given that cognitive science can be generally understood as "the interdisciplinary study of mind and intelligence, embracing philosophy, psychology, artificial intelligence, neuroscience, linguistics, and anthropology," many who research within the field have become increasingly interested in the ecological and evolutionary scaffolding that supports and might even constitute cognitive processes.[4] In subsequent chapters I discuss at great length how this trend has impacted philosophy, but for now suffice it to say that Darwin's impact is felt beyond ethology. Cognitive science, one might argue in a vein similar to Tinbergen's, will not adequately account for cognition unless it addresses the adaptive and evolutionary components of those processes and why we consider them cognitive in the first place.

Griffin saw this connection and applied it to nonhuman animals. If the environment shapes biological development, it stands to reason that it shapes cognitive development as well, and this should be true of any creature we deem to have a mind. Determining what animals have minds is of course a contentious project itself, but assuming humans are not the only ones, it makes sense that at least some animals have rich inner mental lives like humans, perhaps can form beliefs, even have conscious awareness of others' beliefs, and so forth.[5] Thus cognitive ethology was begun as a new research paradigm that would demand animal scientists broach the subject of animal cognition in all its complexity.

Not everyone has welcomed cognitive ethology with open arms, and some researchers are emphatically opposed to its existence. Bekoff and Allen in "Cognitive Ethology: Slayers, Skeptics, and Proponents" (1997) provide a concise summary of those who have spoken out since Griffin introduced his new field of inquiry. I shall not rehearse the many strands of naysaying and support for cognitive ethology here; as we proceed, many of the concerns from critics of Griffin become apparent, and I do my best to address them in turn. The two concerns that I focus on most are folk psychological theorizing and anthropomorphism. Cronin (1992), for example, accuses Griffin of using vague and misleading concepts like "consciousness"—concepts we have yet to satisfactorily prove much about in ourselves—and applying those liberally to animal behavior with little to no evidence that such *folk* concepts are in fact real, let alone that they exist in the animal in question. In a related critique, Humphrey (1977) argues that Griffin's ideas smack of an egregious anthropomorphism, inferring from behaviors in animals that are nothing like us to similar mental states simply because the behaviors are loosely akin to things humans do when they are having this or that thought or conscious experience. These are worries that I take seriously, and I think there are ways to avoid overly vague folk psychology and egregious anthropomorphism while still insisting that animals have thoughts, emotions, and consciousness. In the last section of this chapter, I return to the worry of anthropomorphism in more detail in order to provide a cautionary note about the arguments I make throughout this book. But now, I want to let the discussion "go to the dogs" so we can see how they figure—or, as some might argue, how they do not figure at all—into the cognitive ethological framework Griffin envisioned.

Exploring the Umwelt of the Dog

Cognitive Ethology combines the interdisciplinary aims of cognitive science with the ethological argument that if we want to truly grasp what animals are doing, why they are doing it, and what purpose it might serve, we must examine them as they exist "naturally." That is to say, a proper account of animals must view them in the context of their *umwelten*, or their "lifeworlds."

Jakob von Uexküll ([1934] 1957) was the first person to utilize the term *Umwelt* in connection to animal behavior, and though the German word typically translates as "environment" or "surroundings," von Uexküll emphasized how these environments become specialized "worlds" to the animal. Two animals living in the same physical environment can have vastly different lifeworlds, depending on how those environments get utilized and are perceived by the animal. Hence, another translation of *Umwelt* is "self-world." Von Uexküll's colleague, a semiotician named Thomas Sebeok, argued that the umwelt was one of the "biological foundations that lie at the very epicenter of the study of both communication and signification in the human [and non-human] animal" (1976, ii). To understand what an animal is doing—including what it is communicating and thinking—is therefore to immerse oneself in that animal's umwelt.

The philosopher Giorgio Agamben (2004) went a step further and claimed that the environment itself is not some static *thing* but is rife with "markers of significance"—features and events in the world that matter specifically to that animal in virtue of how the animal relates to those elements and what those elements afford the animal. For example, a certain odor might not mean anything to me, but it might carry extremely important information to my dog (Sokolov et al. 1997; Rodionova et al. 2009), despite the objective fact that both I and my dog are in the presence of the very same volatile chemicals. This is not just a fact about the dog's olfactory anatomy versus my own. Granted, my lack of vomeronasal organs and my much smaller olfactory epithelium are part of the reason why I don't suddenly get overexcited when we walk by a certain bush every day, but the physiology is only one piece of the puzzle. Recalling Tinbergen's four questions, if I really want to know why my dog cares so much about smells that I cannot even detect, I should also consider what those smells are allowing the dog to glean from the environment, how they are helping the dog learn from and adapt to that environment, and how this deep connection between the volatile compounds of the world and my dog's neuroanatomy has shaped *many* canine behaviors before my dog.[6]

This brief discussion of umwelten is intended to shed light on how ethologists tend to conduct their research. Pulling a chimpanzee out of her typical umwelt in the rainforest and conducting experiments in a laboratory not only can skew results, but it will fail to address the myriad ways behavior arises in the context of very specialized semiotic relationships between the chimpanzee and her environment. To be sure, some questions might be better answered in a lab, and if the "natural" environment turns out to be around humans (true of chimps raised in captivity), this criticism will not necessarily apply. In short, we must pay close attention to the questions themselves and to what extent the ecological niche the animal is accustomed to living in might affect the results generated in seeking to address the questions. When you apply this ethological approach to cognitive science, as Griffin envisioned, you are better equipped to successfully answer questions not only about animal behavior but also about animal cognition.[7] And for Griffin, cognition meant more than simply outward behavioral reactions to stimuli, but the inner mental life—consciousness, intentionality, affect, and reason—those capacities that scientists had either willfully ignored or, worse, assumed not to exist in the first place. As I have noted, cognitive science is by its very nature a broad approach to studying "the mind" in all the multifaceted ways that subject can be understood. This includes philosophical perspectives that take seriously the ability of thinking beings to do things like introspect, reflect, have qualitative aspects to their experiences, and feel emotions. Thus cognitive ethology should ideally allow us to peer inside the minds of animals without forcing them into situations that are artificial and ethically questionable.

This brings us to the question about domesticated animals, and specifically *Canis familiaris*, the dogs with whom so many humans share their homes. In the last decade or so, the world of animal behavior has been flush with studies and findings pertaining to canine behavior and cognition. Researchers claim to have demonstrated that dogs can sniff out cancer in humans (Willis et al. 2004; Cornu et al. 2011; Ehmann et al. 2012), engage in inferential reasoning (Brauer et al. 2006; Erdőhegyi et al. 2007; Aust et al. 2008), detect seizures (Edney 1993),[8] tell time (Warren 2013, 2015; Horowitz 2016), get depressed and even suicidal (Braitman 2014), and love humans as deeply as humans love them (Odendaal and Meintjes 2003; McConnell 2006; Berns et al. 2015), and that having a dog can prevent heart disease and prolong the life of the human guardian (Le Roux and Kemp 2009; Mubanga et al. 2017). I discuss many of these findings in subsequent chapters, and I do so with a skeptical lens. While I am fascinated by the abilities of dogs, I am also keenly aware of the ease with which researchers and philosophers alike can overlook confounding variables and help themselves to conclusions based on little more than wishful thinking. Nonetheless, it is undeniable that dogs have been more in the spotlight than ever because they have some unique capacities that scientists have largely overlooked until now. The findings I am most interested in examining are those that suggest dogs' cognitive capacities have been grossly underestimated. Researchers like Brian Hare, who created the website Dognition for collecting data from dogs in their own homes as their owners/guardians play games with them, and Ádám Miklósi, founder of the Family Dog Research project in Budapest, have forged a path in cognitive science and animal behavior that has resulted in dogs being accepted as genuinely worthy of rigorous scientific study.

But how do dogs figure into the cognitive ethological framework I have discussed so far in this chapter? It might be argued simply that they don't. That is, dogs, much like other domesticated species such as *Felis catus*, the common housecat, are always already improper subjects of any ethological study, cognitive or otherwise, precisely because they are living with humans and are impossible to study in any sort of natural habitat. An easy rejoinder is that, for many dogs, their umwelt *just is* in the homes of humans. Even street dogs in India—although they do not sleep in beds with humans the way many dogs in U.S. households do—are immersed in a human world. Russian street dogs have capitalized on this feature of their umwelt and have learned to utilize the Moscow metro to travel to food-dense locations (Boyd 2016). Thus I think the right approach when studying dogs is often, necessarily, going to involve studying them *with humans*. However, it is worth pausing for a moment to consider why ethologists and other scientists worry about human interference in animal studies.

It would be rare to find a book or article that grapples with questions concerning the cognitive capacities of animals and human interference that does not discuss the Clever Hans Phenomenon. Hans, a horse trained by a mathematics instructor, Wilhelm von Osten, in Berlin in the early twentieth century, made his debut after several years working with his owner. Osten claimed that his horse could perform arithmetic operations, tell time from a clock, count the number of persons in the room, and recognize and identify playing cards. Sure enough, when asked to do so, Hans would perform these tasks with astonishing accuracy. Many people, especially among the scientific community, were skeptical and assumed this all must be some elaborate hoax, but after the German Board of Education set up and executed a period of testing that lasted more than a year, it was concluded that there was no hoax—Hans really was that clever. That is, until Oskar Pfungst, a professor of biology and psychology, discovered and reported (1911) that Hans could not answer questions if the person asking him did not know the answer. So, for example, if Hans was asked to identify a composer based on the song being played, if the person asking him the question did not know the answer, Hans would fail to pick out the correct composer from a list of possible answers as well. Moreover, when Pfungst placed a screen between the questioner and Hans, such that Hans could not see the face of the person asking him a question, he was unable to answer correctly. As it turned out, Hans could answer so many questions correctly because he was reading subtle cues from the faces of his questioners (Samhita and Gross 2013). Thus, just as quickly as Hans had risen to equine fame for being the smartest animal alive, he was stripped of his title, and he is now remembered instead as a warning for ethologists. Today it is commonplace wisdom among ethologists that the animals being studied should never be able to see human faces. This rule is often broken, however, and Clever Hans is discussed in relation to canine studies as we proceed.

The Clever Hans Phenomenon now serves as a reminder that if you are not careful when setting up ethological studies, you could very well measure something entirely different from what you originally intended. Moreover, the Clever Hans case demonstrates the need to be especially careful not to prematurely claim evidence against the *null hypothesis*—a fact or set of facts taken to be true before beginning the experiment. Osten and others thought Hans's behaviors justified rejecting the null hypothesis that horses should not be able to engage in mathematical reasoning, but in fact, upon closer inspection, it was shown that this hypothesis was not disproven at all. Osten was completely unaware that he was providing the clues to Hans, so the Clever Hans Phenomenon was not an intentional hoax, but it was certainly not what it appeared to be prima facie. Humans are notoriously complex and varied in how they express themselves facially, particularly when it comes to affective gestures (Du et al. 2014). But we are often not aware that our faces are giving away any information whatsoever. Consider the way an experienced poker player may wear dark glasses to avoid giving other players a "tell"—a very subtle facial gesture that conveys that the person is pleased or displeased with the current hand. Similarly, Osten and all the other persons who interacted with Hans when he answered correctly were accidentally revealing their "tell"—by glancing for a millisecond in the direction of the correct card, curling their lips just so when the correct number appeared, or some comparably indirect gesture. Then when Hans or any other animal under investigation performs tasks that supposedly only humans can master, it is easy to get prematurely excited and publish papers that proclaim incontrovertible proof that animals think just like humans. Instead, many ethologists argue, Clever Hans reminds us how problematic studying animals can be, especially if proper scientific rigor is not applied.

But there is another interpretation to the Clever Hans story that gets much less consideration. While Hans might not have provided any evidence that horses can perform arithmetic or recall famous composers based on musical samples, he did something quite clever indeed. He observed the almost undetectable facial gestures of humans and used them to perform the requisite hoof-tapping that would grant him rewards. We might even go so far as to say Hans was *better* at reading human faces than humans are. After all, it took quite a long time for Pfungst and his colleagues to figure out how Hans was getting so many answers correct. This last suggestion is of course an example of unwarranted anthropomorphism, but the claim that horses are adept at reading human facial cues is not an untestable hypothesis, and it has been the subject of many studies recently. So far, horses, rhesus monkeys, and dogs are the only nonhuman animals shown to track human faces *the same way* that humans track human faces.[9] Why exactly this is, what purpose it might serve, and what this capacity means in terms of a horse or dog possessing a "theory of mind" are as yet unanswered questions. But the point is that Hans, albeit not a mathematician, was nonetheless a smart horse, and one who inspired a lot more research.

This brings us back to dogs. If the standard ethological charge is true—that human interaction will always nullify results in nonhuman animal studies—then most ethological studies of dogs are doomed before they can even begin. After all, the umwelt of dogs almost always includes humans, and not just as an occasional passing interaction. There are now more dogs than children in the United States. Approximately 44 percent of U.S. households have dogs, with a canine population estimated in 2016 at 78 million and rising (APPA 2018), and those dogs are increasingly being treated like the human children they outnumber. According to the same national pet owners survey, 68 percent of dog owners claim to give their dogs Christmas presents every year. Dogs sleep in the bed with us, eat our table scraps, and sit on the sofa with us while we watch movies. The little spaces we have carved out and call our own—our houses and yards, apartments, cars—are shared with canines just as much as with other humans. The dog's ecological niche, in other words, is ours. And it is us.

Developing an ethological study of dog cognition that avoids the pitfalls of human tampering would be nearly impossible, unless of course those studies were designed to be carried out on wild dogs. If we really want to test to what extent most domesticated dogs can think and carry out problem-solving tasks, there is no way to do so without accepting that as domesticated animals who also live in extremely close proximity to humans, dogs will always be "spoiled" by our influence. Of course, there are ways to study dog behavior independent of human interaction for small periods of time.

As we proceed, I survey some studies that attempt to do this. However, unless the dogs in question are in fact wild—not just feral, as many "street dogs" interact with humans quite a lot—domestication means that the phylogeny and ontogeny of the dog has been shaped and informed by humans. Domesticated dogs, it might seem, are antithetical to the overarching aims of ethology.

While I am sympathetic to some important differences between studying, say, chimpanzees as they interact with other chimpanzees in the rainforest and studying domesticated animals like dogs, I do not think that ethology must exclude the latter from its purview. Several compelling reasons convince me that canine ethology not only is possible but is an important and needed branch within this field. First, as I mentioned above, what arguably counts as a natural habitat for the dog just is an environment that includes humans, their homes, technologies, and all the other interactions that have emerged over the last fifteen thousand years. This point rests on the term "natural" and how it is construed. To philosophers, the debate over the meaning and usefulness of the term "natural" is likely well known, so readers familiar with these issues can skip the next section. For those less aware of this philosophical tension, I provide a brief rehearsal of why I think it is not so easy to stamp the label "natural" onto some interactions among animals and their particular ecological niches while simultaneously referring to dogs or other domesticated animals as "unnatural" simply because they are "contaminated" by the human umwelt.

The Umwelt of Domestication and a False Choice between Natural and Unnatural

What does it mean to insist that animals be studied in their *natural* habitats and why is it important to many ethologists that this standard be upheld? It seems that the simplest way to understand what is meant by the term "natural" here is in terms of place: a natural habitat of a species is one the animals occupy of their own accord, without human intervention. Specifically, "natural" is *not* in the laboratory or in the zoo. As I explained above, this stipulation is not an arbitrary one, as we know that artificial settings such as labs and zoos can alter experimental methodologies and findings. What a dolphin does in the ocean in response to a stimulus is perhaps quite different from what that same dolphin would do if presented with that same stimulus while in a tank at an aquarium. What's more, the dolphin in the tank is likely to have interacted with or been trained by humans, and these interactions shape how the dolphin will respond to various stimuli as well. The argument that animals behave and think in ways that are directly related to and perhaps even exclusively enmeshed within the environments they naturally occupy harks back to the writings of John Dewey: "The idea of environment is a necessity to the idea of organism, and with the conception of environment comes the impossibility of considering psychical life as an individual, isolated thing developing in a vacuum" (1884, 56–57).

What Dewey is saying is that to study animals is always already to study the environment in which they live. I could not agree more wholeheartedly. In fact, as we proceed, this notion of a coupled system consisting of organism + environment—as opposed to just the organism itself—being the real subject of interest in ethology is shown to compare favorably with what has now been dubbed "ecological cognition" in the philosophy of cognitive science (Hutchins 2000). A mind, so the theory goes, is always tied to its particular ecological niche, and that symbiotic system of organism + environment is what constitutes this or that cognitive process. I return to this idea later, but for now let us examine in a bit more detail where the assumption that animals are best studied *together* with their environments, and more specifically their "natural" environments, leads us.

To be sure, a laboratory or aquarium is a quite different setting in which to approach questions concerning the cognitive aptitude of animals, but what about the homes of humans? This is the chief place one would expect to find dogs, at least in many parts of the world. Should dogs be excluded from ethology because their environments are inherently human-filled and therefore unnatural? It has long been thought that domestication is a process that corrupts the natural organism + environment system because it introduces the human element into an otherwise human-free ecology. Moreover, the standard account of domestication holds that humans deliberately sought out wild animals, such as wolves or boar, and transformed them into creatures that could help hunt and provide companionship, or be farmed and turned into food. The way dogs or pigs on farms behave is thus vastly different from the way their wild ancestors did, precisely because they are raised in completely different environments and are not afforded the same ecological interactions they would have had if they were left to exist apart from humans.

Whether we think of this domestication process as human guided or animal initiated is beside the point, though that discussion is worth having independently. The more pressing question I asked at the beginning of this section was whether domestication—however it occurred—entails rendering a genuine ethological study of that animal impossible because the animal is no longer living in a human-free, natural environment. I think the answer is, at best, far from clearly settled. First, to say that an environment filled with humans is unnatural is a bizarre claim, as it is tantamount to arguing that humans are somehow outside of nature. Of course, when people use the term "nature," they often do so in direct opposition to typically human-infested territories like cities and suburbs. But this distinction is arbitrary, because we often say things like "I need to escape into nature for a few days" and then load our cars with coolers full of genetically modified food, head out into the woods and set up camp under a polyester tent, make a fire, cook the food, and celebrate our escape with a glass of wine. We have "infiltrated" nature for the moment, only to return to our artificial lives in the city. But our lives in the city are just as natural, if not more so, than our hiatus camping in the woods. Most of us are quite *unnatural* at camping, and the mere thought of living without electricity or in the company of bears and snakes terrifies us. Modern-day humans, in other words, are arguably most natural and in their natural habitat with the vast array of tools, buildings, and sociological structures that tend to define cities and towns. This city-versus-wilderness model of unnatural-versus-natural is just one of many examples of the ways we don't have a particularly solid account of what precisely makes something natural or unnatural in the first place.

But what of the interaction between humans and animals, as in the case of domestication? If it is true, as I argued, that being filled with humans is not enough to qualify a space as an unnatural one, does it perhaps become so when we add in the fact that many of these human-infested spaces are also replete with animals and, moreover, that these animals live with, have been trained by, and are dependent on humans for their survival? The cute Pomeranian sitting on the lap of a woman in her fifty-story apartment building in Chicago is the result of domestication par excellence. Surely this dog is living "unnaturally" compared to his noble wolf cousins who live in dens in mountain outcroppings and hunt in packs. Indeed, the behaviors of this Pomeranian and the wolves are going to be vastly different, but then so are their phylogenetic lineages. The domestication process itself is what has made the Pomeranian possible, and the existence of dogs as we know them is entirely dependent on interaction over the course of thousands of years. Domestication, you might say, is as natural to *Canis familiaris* as tool use is to *Homo sapiens*.

The quibble over the terms "natural" and "unnatural" as it pertains to domestication could be just that: a semantic dispute. Perhaps it doesn't matter if we label domesticated dogs as natural or not, because what is really at issue, the ethologist might claim, is whether, in consequence of that domestication process, the dog cannot be understood free from the influence of humans. Conducting studies free from human interference is, after all, one of the baseline edicts of ethology. In other words, we might grant that the debate over whether we can properly label the cohabitation of dogs and humans a "natural" process is an intractable one. We could likewise argue all day over what constitutes a "natural human"—does tooth brushing disqualify you from that category, and what about wearing clothes? What is really troubling to ethologists about animals like dogs is not so much whether domestication is a natural process, but whether that process has somehow altered the animal in such a way that the animal does not think or behave the way it would without human interaction.

Taking care to avoid misinterpreting animal behaviors—especially if they are the result of human interaction—is indisputably crucial. A repeat Clever Hans story is never desirable for anyone studying animal cognition. However, it is not logical to conclude that simply because an animal has interacted with humans, its behaviors are no longer worthy of serious investigation. The fact that dogs are so good at reading human gestures and facial expressions undoubtedly has something to do with the interactions between humans and dogs, not just in the short term—the lifespan of one dog living with human(s)—but also in the more extended relationship between the two species over thousands of years. As it turns out, dogs have some impressive abilities to extract data from humans that very few other animal species have heretofore exhibited. And those abilities have helped the dog survive and proliferate.

Furthermore, the capacities an animal might develop from interacting with a human could indeed be considered intelligent, especially if by "intelligent" we mean highly adaptive to changing environments and capable of solving problems in a diverse array of scenarios. Even Clever Hans, though he was not "clever" in the way his handler claimed he was, was nonetheless skilled at reading the subtle cues present on human faces. And this was something he learned to do *without* training—in other words, without overt instruction by a human. We now have compelling evidence that dogs similarly excel at detecting facial cues (Guo et al. 2009). And as with horses, dogs don't need to be taught to do this. It is something that seems to have been incorporated into the species over the course of its coexistence with humans. It is relatively unsurprising that both horses and dogs have this capacity, given how closely both animals have traditionally worked, played, and lived with humans.

What I have discussed so far regarding the naturalness or unnaturalness of domestication and whether interaction with humans entails that an animal is not worthy of consideration under a cognitive ethological study leads me to posit two theses:

1. It is worthwhile to regard dogs and other domesticated animals as genuine subjects of cognitive ethology, especially if we allow that the "natural environment" of *Canis familiaris* always already includes humans.
2. Intelligence is a complex concept, and it is not at all settled— not in philosophy nor in cognitive science—what the proper understanding of it should be.

I think I have provided sufficient evidence in this chapter already for the first, although I return to it at various points throughout the book. As to the second, I delve more deeply into this claim as I work through chapters 2 and 3, where I discuss recent findings in canine science, what they seem to imply about canine intelligence, and how affective cognition plays into all of this. For instance, if there is evidence that dogs are highly skilled at recognizing facial expressions on humans and coding them in terms of emotional meaning, should we count this as a type of intelligence? The division between emotion and rationality is a long-standing dichotomy that has marked philosophical discourse as well as a host of other disciplines. As I argue in chapter 2, however, this assumed divide has been eroding over the last couple decades when it comes to thinking of "emotional intelligence" in humans and the ways in which affect undergirds even the most basic cognitive capacities. So, if we are willing to extend the purview of cognitive science to include affective processes in what is more generally considered cognition, then it is arguable that we should approach nonhuman animals who exhibit similar affective skills with the same open-mindedness.

At this point it is probably clear that I have my own biases, so I should take a moment to note that I am in fact convinced that many nonhuman animals think, experience emotions, perform inferences, and perhaps even possess a sense of right and wrong. All the same, as a philosopher who is inherently skeptical and bound to exhaust all possible doubts, I have made it a top priority to approach every experimental finding that purports to demonstrate any of these abilities with a critical lens. When it comes to canine intelligence, the central focus of this book, I am even more cautious, partly because the study of the canine mind is so new and there are bound to be some missteps as the field matures, and partly because given how closely dogs have lived with humans for the last fifteen thousand years—indeed, given how closely I myself have lived with dogs for my whole life—it is especially easy to fall into the logical trap of wishful thinking regarding the abilities our canine companions seem to possess. This is particularly the case when we are confronted with news stories or popular scientific publications with click-bait titles such as "Brain Scans Reveal What Dogs Really Think About Us" (Fisher 2020), thereby gaining an immediate sense of validation for what we have intuitively felt to be true of our dogs but have had no scientific evidence to support. I accordingly proceed with great caution so that I might be as philosophically and scientifically honest as possible. This means not engaging in unwarranted anthropomorphism, wishful thinking, or willful ignoring of countervailing data.

While all of this is of utmost importance to me, I want to end this chapter with a small section that considers the opposite worry, namely what Bekoff and Allen (1997) have termed the "slayer" approach to ethology. Avoiding wanton anthropomorphism is advisable, to be sure, but so too is avoiding a stubborn and species-chauvinistic anthropocentrism—or, worse, *anthropectomy*, the denial that animals have any humanlike traits whatsoever. As is the case with many issues in philosophy and science, dogmatic adherence to extremes is likely to lead to indefensible theories. The question regarding anthropomorphism and its place in a proper science of animal minds is not a simple one and likely does not have a simple answer.

A Cautionary Note about Anthropo-anything

Conwy Lloyd Morgan, one of the founders of comparative psychology, is perhaps best known for a dictum pertaining to the appropriate study of animal behavior that has had incredible staying power over the years. Known as Morgan's Canon, the decree states, quite simply, that "in no case may we interpret an action as the outcome of the exercise of a higher psychical faculty, if it can be interpreted as the outcome of the exercise of one which stands lower in the psychological scale" ([1894] 1977, 53). Morgan's Canon has been widely interpreted to be an attack on anthropomorphism because it appears to suggest that humans are often guilty of observing in themselves "higher psychical faculties" that lead to certain behaviors, and then ascribing those same faculties to animals when similar behaviors are observed. This dubious double-induction process, according to Morgan, is a consequence of human introspection coupled with an overly liberal attribution of similitude between our behaviors and those of other animals. He warns that both acts of induction should be executed with great care so as not to make mistakes in ascribing mental processes to behaviors when there is insufficient evidence that those mental processes obtain.

Morgan's Canon has led animal researchers such as Clive Wynne (2007) to argue vehemently against an uncritical anthropomorphism that Wynne takes to be the hallmark of much of ethology, because it delegitimizes findings in animal behavior. He discusses how easy it is for him to label certain behaviors his dog exhibits as "remorseful" and surmises that this anthropomorphic tendency is a conditioned response—that is, over time we become accustomed to inferring that like behavior equates to like mental states, and without the presence of any defeater, we are "rewarded" in the sense that our inference provides a decent baseline explanation. However, humans seem to be innately endowed with the capacity to infer mental states in other humans based on behavioral similitude,[10] so what Wynne sees as a conditioned response might be even more strongly hardwired into us. In short, we cannot help but see behaviors that are in any way similar to our own as evidence for a comparable inner mental life. Overdoing it, though, is where we run into trouble.

Gordon Burghardt (2007) argues that Wynne commits the nominalist fallacy, assuming wrongly that naming something suffices to explain it—and, in the case of anthropomorphism, dismiss it. Ironically, Wynne, like many ethologists, is wary of nominalism because it pushes off the task of explanation onto a name. Calling a behavior "predatory" does not *explain* the behavior, especially not according to Tinbergen's four criteria. So it is important not to slip into a naïve nominalism when cataloguing animal behaviors. The goal is to explain behaviors, from the ontogenetic and phylogenetic underpinnings to the ecological and functional roles those behaviors play. However, when Wynne asserts that anthropomorphism is guilty of the nominalist fallacy—that is, by anthropomorphizing, we are simply naming behaviors as "mentalistic" and not explaining those behaviors—he is committing his own nominalist fallacy because he is labeling a behavior, namely human labeling, as explanatory of that behavior. Furthermore, he goes on to assert that anthropomorphic labeling is to be dismissed because the behavior has no genuine explanation.

Burghardt argues instead that we adopt a "critical anthropomorphism," which allows for the fact that it is difficult, if not impossible, to avoid seeing the world in terms of human characteristics. He accuses Wynne's argument of being uncritically anthropocentric and of regressing to a Skinnerian behaviorism and naïve objectivism. It is inevitable that we will anthropomorphize, he claims, and the trick is not to dismiss it but to be critical and measured in how we do so. I quote Burghardt at length:

> Not only is critical anthropomorphism useful in developing hypotheses, an unreflective objectivism is bad science: in this case anthropomorphism by omission, an idea developed in another essay cited by Wynne but misrepresented (Rivas & Burghardt, 2002). By dismissing our own status as animals evolved to deal with the problems of living that other species also have to face, and attempting to be completely objective, we fall into serious errors as readily as through being naively anthropomorphic. (2007, 137)

Here Burghardt links "unreflective objectivism" to "anthropomorphism by omission," the failure to appreciate that an animal's umwelt can be and almost always is vastly different from our own. We assume that the human perspective is *the* perspective in the world and wantonly attribute it to other species. Attributing humanlike qualities to other animals in this sense is therefore, rather strangely, *both* anthropomorphic and anthropocentric. As Rivas and Burghardt note in their 2002 paper, it is not enough to denounce anthropomorphism and suppose that this dismissal suffices to make the research entirely objective. Not only is this an instance of the nominalist fallacy, but it also fails to address the fact that in so doing, we assume it is even possible to take up an entirely objective viewpoint, freed entirely from our own subjective consciousness and all the human biases inherent therein.

Burghardt's suggestion that we adopt a "critical anthropomorphism" asks us to recognize our tendency to see the world through our specifically human eyes, while also working to appreciate the uniquely situated perspective of other animals. Rivas and Burghardt speak of putting themselves in the animal's situation as much as possible, which echoes the ideas of Timberlake and Delamater (1991), who claim that not only should we aim to place ourselves in the "shoes" of the other animal, but we should attempt to walk in them. The proposal that we adopt critical anthropomorphism appears in various forms abundantly throughout the ethological literature. Frans de Waal (2006) argues that it can provide a useful tool for investigating the extent to which animals think and act similarly and dissimilarly to humans, and Alexandra Horowitz (2010) utilizes this approach in her canine research, where she quite literally tries to "be a dog" by moving about the world on all fours and attempting to rely more on olfaction than on sight to perceive her surroundings. The idea is that when engaging in any comparative psychological pursuit, at a minimum, two thinking beings must be compared. Humans are the ones doing the investigation, presumably, to learn about the minds of other animals, and so it stands to reason that we are a good point of comparison. In adopting a "critical" anthropomorphism, however, it is imperative that we take the animal's specific ecological and evolutionary contexts into consideration and realize that no matter how comparable our behaviors might appear, we don't occupy the same world, at least not entirely, and we will never truly be able to "get inside" the minds of other animals (Daston and Mitman 2005).

Although philosophers have not paid nearly as much attention to anthropomorphism and anthropocentrism as have ethologists, there have been some discussions in the philosophy of mind that are worth noting here, as I think they shed important light on my view of the best way to address the issues for the purposes of this book. Thomas Nagel, in perhaps his most famous article, "What Is It Like to Be a Bat?" (1974), argues for an approach to understanding the mind and consciousness that resembles critical anthropomorphism. By asking us to consider what it might be like to be a bat, Nagel concludes that although we can supply some objective facts about bat life—that they fly, echolocate, and so forth—we will never be able to capture the *subjective* nature of that experience. Even if I attempt to "walk in the bat's shoes," as some of the ethologists mentioned in this chapter urge researchers to do, at best I will become more sympathetic to the ways in which bats navigate their world. Take the case of Daniel Kish, who in infancy lost his eyes to retinal cancer. Over the years Kish has developed the ability to obtain information about his environment by making clicking sounds and noticing the particular auditory qualities as those sounds bounce off objects nearby (Arnott et al. 2013). Nagel would say that Kish is able to simulate being a bat way better than most sighted persons can, and indeed better than many non-sighted persons as well, since the blind tend to rely most heavily on tactile perception and haptic feedback to navigate the world. Nevertheless, Kish does not know what it's like to be a bat because, quite simply, he only knows how to be himself, a human who happens to utilize echolocation. To know what it is like to be someone, in other words, is *to be that being*.

One of the major concerns Nagel is addressing here dates back at least as far as Descartes, and arguably further, though it was Descartes who famously introduced it explicitly into the philosophical canon.[11] The issue, aptly dubbed "the problem of other minds," refers in its most general sense to the epistemological and conceptual difficulties in properly understanding any thinking beings that are beyond our own subjective minds. I shall not delve into the nuanced and varied approaches to this problem, but will just take a moment to discuss how it figures into the ethological debates concerning anthropomorphism and anthropocentrism. Descartes's inquiries into the nature of thought led him to adopt a radically skeptical and wholly egocentric stance regarding the world. By doubting anything and everything that was not absolutely certain, he was left only with his own subjective conscious experience, and the conviction that he, himself, was thinking. This alone could not be doubted, and from this starting point he began to try to build more knowledge. Unfortunately, for Descartes this meant that proving anything about how other beings think was rather tricky business. Indeed, it was difficult for Descartes to prove that anyone but himself thinks. This skepticism about other minds—whether in doubting that they exist at all or doubting that we can ever know entirely what that thinking is like—is known as *solipsism*. The most radical form of solipsism—the belief that nothing beyond my own mind thinks or even exists—is not taken seriously in philosophy, for obvious reasons. *Methodological solipsism*, on the other hand, is a position that is heartily endorsed. Adopting a framework that is solipsistic, but only methodologically, means I recognize that the problem of other minds is intractable—I will never get indisputable proof of *what* other beings are thinking. I can, however, reasonably conclude that other beings *do* think, and can therefore approach the questions of what those thoughts are about, how they are constituted, and so forth from the standpoint of seeking as much inductive certainty as possible. Jerry Fodor (1980) famously makes the case for methodological solipsism in philosophy of mind, and importantly notes that we must face up to the problem of other *human* minds just as much as nonhuman minds. Philosophically, methodological solipsism is hard to refute, as it recognizes and is based on another irrefutable fact of studying other minds: we can only get so far. Given our limitations, what can we hope to infer about other minds, then, human and nonhuman? This is where taking methodological solipsism together with the critical anthropomorphic stance defended by Burghardt and others seems a promising route.

The doubly inductive process by which we infer that other humans have thoughts like our own—because, first, we know we have thoughts and we observe those thoughts in conjunction with behaviors of our own, and then we see those same behaviors in others and conclude they have similar mental states—is the same process by which we suppose nonhuman animals might have thoughts like ours as well. To be sure, the double induction is even more problematic when comparing human and nonhuman thoughts and behaviors. In view of the physiological and ecological differences, a quick comparison is not so easy. Nonetheless, this is the same problem that plagues philosophers attempting to explain subjective consciousness in other humans. Indeed, Nagel's famous article was not really about bat minds at all, and instead focused on human consciousness and how objective science will almost assuredly never fully capture a phenomenon that is, by its very nature, subjective.

Ultimately, Nagel argues that an *objective phenomenology* is needed to address the problem of other minds, which might seem counterintuitive, given that phenomenology is a tradition in philosophy steeped in subjectivity. Yet the father of phenomenology, Edmund Husserl, was clear that by engaging in careful descriptions of experience, he was aiming to explain not just his own subjective consciousness but consciousness as such (1960, 1970). As Gallagher and Zahavi (2007) also note about the phenomenological method, there is a big difference between providing a subjective account of experience and an account of subjective experience. Phenomenology, as they understand it, is supposed to be the latter.[12] Thus Nagel is not claiming that an account of other minds will be comprehensive; no objective science could ever provide that. Instead, what he suggests is that in developing a phenomenology of consciousness, we do so in a way that makes our descriptions accessible to others who might not have the same experiential base. For instance, if I am trying to explain what it's like to skydive, it will help if I find analogous experiences that persons who have never done it can better relate to. Again, this will never be a perfect solution. Unless those persons go skydiving, they will never know what it's like from a subjective standpoint. And even if they do go skydiving themselves, they won't know what it's like *for me* to go skydiving, and similarly, I will never know exactly what it was like for them to finally try it. Still, the analogies we draw in trying to explain subjective experience to one another are not useless. In fact, in Nagel's view, these analogies are the most crucial tools we have for conveying our experiences. Scientific facts about the physiological causes of thoughts or the phylogenetic underpinnings of consciousness will certainly help us in determining *how* human thought arises, but it is objective phenomenology that will make those accounts much more personally relatable.

Nagel's "objective phenomenology" echoes quite a bit of the methodology urged by Dan Dennett (2003c) when he says we must engage in *heterophenomenology*.[13] Dennett, however, takes a much more eliminativist view about the supposedly special properties of inner mental life—what philosophers refer to as *qualia*, or the "raw feels" of consciousness. This is not to say Dennett does not believe in consciousness; he just redefines it as "a bag of tricks" (2003a, 2003b). More important, as he points out, even our own subjective access to the inner workings of our minds is not incorrigible. The phenomenon of change-blindness explored by John Grimes (1996) demonstrates how very little we are consciously aware of in our immediate perceptual environment, even when visible changes are taking place right before our eyes. All of this suggests to Dennett that the best way to tackle questions concerning conscious experience is to engage in a phenomenology that acknowledges the importance and necessity of the third-person perspective—"phenomenology *of another* not oneself" (2003c, 19). We can learn a lot about the minds of others, and about our own subjective consciousness, from objective, third-person accounts, many of which are generated from psychological and neurobiological studies. Though much confusion surrounds heterophenomenology, especially with regard to a third-person account of a very first-person phenomenon, I think Dennett is correct in arguing that this is as close as we are going to get to meeting Husserl's demands to describe consciousness as such. The only first-person account I can accurately give of experience is my own. Anytime I seek to describe the experiences of others, I do so from my own subjective vantage point, and thus my account of *their* mental states is always already objective. Hence, Dennett claims, heterophenomenology is "the sound way to take the *first* person point of view as seriously as it can be taken" (2003c, 19).

To tie this philosophical digression more closely to the previous discussion of anthropomorphism and anthropocentrism, I think both Nagel and Dennett argue for something along the same lines as Burghardt, Waal, and others who advocate a "critical anthropomorphism." Recognizing the impossibility of seeing other animal behavior from any perspective but our own, while also making a genuine concerted effort to take into consideration that animal's vastly different umwelt, allows the most headway to be made in cognitive ethology, as I see it. Much as I will never really know what it's like to be another human, I will never know what it's like to be a dog. But just as imperfect knowledge of other human minds does not stop me from gaining a pretty solid account of how human minds work generally, it should not dissuade us from trying to make sense of our nonhuman neighbors, even those who seem to be worlds apart from us. Peter Godfrey-Smith takes up this challenge in his recent book *Other Minds* (2016), when he attempts to understand octopi minds. Like Dennett's, his methodology is, of necessity, based in third-person, objective science. All the same, in trying to get a sense of what it might be like to think like an octopus, a certain amount of anthropomorphism is not only unavoidable but highly useful. Without relating octopi experience to human experience in any way, it seems unlikely that we could ever begin to appreciate how incredibly *different* those experiences must be.

The approach I take throughout the remainder of this book, then, is aligned to some extent with the heterophenomenology of Dennett, along with the critical anthropomorphism argued for by ethologists like Burghardt. Since the book is devoted to exploring the canine mind and its coaction with human minds, I think the "critical" part of the equation needs to be stressed perhaps even more than it would be with a study of, say, octopi minds. So many dogs are inextricably enmeshed in human life that it is easy to conflate their umwelten with our own. Indeed, as I suggested above, their lifeworlds are to a large extent the same as our own. Even street dogs share our umwelt to a large extent and, unsurprisingly, perform similarly on many of the same tests as companion dogs (Bhattacharjee et al. 2017). But even though we might live in the same objective spaces, in the same homes with the same furniture and all the same routines, there are important differences in how the dogs' umwelten are experienced *by the dogs*, in large part because they are physiologically quite different from us. Having evolved alongside us for so many years, they have adapted to us by learning how to do things the way we do them, and it is intuitively plausible that the mental capacities they have reflect our own in many ways because of this unique bond. But this deep entanglement is even more reason to proceed cautiously so as not to engage in uncritical anthropomorphism or unwarranted anthropocentrism. In addition to exploring how dogs think in relation to humans, one of the chief aims of this book is also to examine how human cognition has been shaped and is continuing to change with dogs alongside it. To this end, I do not assume that dogs are the product of an ingenious human species that carefully molded them from wolves into the creatures with whom we now share our homes. This assumption is an example of what I take to be an unwarranted anthropocentrism. Instead, I want to proceed in a critically anthropomorphic manner while also working to develop a critically *cynomorphic* stance. In other words, as Alexandra Horowitz (2016) has recently argued, it is not enough to simply look at a dog and think about its behavior with a critically anthropomorphic lens; we must try to *be dogs*. This does not mean we can overcome the problem of other minds. I will never know what it's like to be a dog so long as I am a human. However, trying to immerse myself in the dog's lifeworld—asking Tinbergen's four questions, but *from the perspective of a dog*—will be useful insofar as it will help to remove as much as possible my own bias qua human. Moreover, as another aim of this book is to determine to what extent dogs have shaped us, taking a cynomorphic stance will better equip us to address the question *What are the ways in which the canine umwelt might shape human cognition?* In the next chapter I begin to seek answers to this question by examining two parallel revolutions, one in philosophy of cognitive science and the other in cognitive ethology. The first concerns the shift away from an internalist and computational account of human thought, and the second concerns a recent explosion of experimental findings pertaining to dogs. As I have argued in this chapter, that latter revolution should be properly considered an ethological one.

CHAPTER 2

Thinking-with Dogs and Dismantling Standard Cognitive Science

Dogs at the Crossroads of Philosophy and Cognitive Science

In the first chapter, I laid a foundation for the overarching methodology I plan to employ—namely, a *critical anthropomorphism* coupled with something quite similar to Dennett's *heterophenomenology*. Given the impossibility of knowing firsthand what it's like to be a dog, we must accept that any account of the dog's cognitive processes and experiences will be third-person. Furthermore, try as we might, we cannot entirely avoid anthropomorphism, for even the most stringent behaviorist must resort to human language to describe and explain behaviors. The task is therefore to determine which explanations have the most evidentiary support. As I suggested at the end of the last chapter, one way we might go about doing this is to try to place ourselves in the dog's "mental shoes"—what I termed *critical cynomorphism*. Critical, because we are aware, thanks to methodological solipsism, that truly taking up the dog's perspective is impossible. We are interested not only in how the dog's behaviors appear to us as they occur in our human umwelt, but also in how the dog's umwelt shapes and is shaped by the dog's experiences. So we must be careful to ensure that any explanation offered takes into consideration the dog's physiological comportment, ontogenetic and phylogenetic background, and the particularities of its umwelt. This critically cynomorphic stance is akin to the methodology Alexandra Horowitz takes up in her 2016 book devoted to decoding how a dog perceives the world via olfaction.

It should be noted that besides arguing that heterophenomenology is the best way to tackle questions concerning consciousness, Dennett is of course also known for arguing that we take up the *intentional stance* when trying to explain behaviors and cognition. Much the way David Marr's ([1982] 2010) three levels of explanation work—we can explain something in terms of its physical organization (the *implementational* level), the way a function within that system achieves a goal (the *algorithmic* level), and the goal of the system itself (the *computational* level)—Dennett thinks there are three types of explanation available to philosophers wishing to explain cognition. When we take up the *physical stance* toward something, we are interested in what it is made of and how the material of the organism or object helps us understand and predict its behavior. For example, the fact that pigeons' eyes look fixedly ahead explains why pigeons bob their heads constantly, unlike humans, whose eyes can dart about in a quick jerky movement called a saccade, allowing their heads to remain still when, say, reading or looking at an object. When we take up the *design stance*, we are trying to understand the behavior of something in terms of its function. What something was "designed to do" might simply mean "evolved to do" if we are talking about biological organisms. A greylag goose, for instance, will use her beak to roll misplaced eggs back toward the nest and will do this with any object that resembles an egg, even only slightly. This behavior can be described in terms of design—that is, the egg-rolling is a motor pattern designed to be activated by eggs or egg-like stimuli. Last, taking up the *intentional stance* means ascribing thoughts, beliefs, and desires to the system to explain the system's behavior. So, for example, if we wish to understand why Paul stopped just short of walking into a bar and went back to his car, only to return several minutes later, and we also discover that his wallet was in his car when he almost entered the bar the first time, it makes sense to say he realized that he had left his wallet in the car, believed that he needed it in order to patronize the establishment, and desired to get a beer, so decided to retrieve his wallet where he recalled having left it. Dennett claims that if the ascribing of intentionality to a system proves useful in predicting and explaining behavior, then we ought to take up that stance. This potentially includes how we talk about nonhuman systems such as other animals, and even machines.

With these methodological concerns further clarified, the purpose of this chapter will be to examine the ways scientists are probing the canine mind and what these findings reveal about the inextricable link between dogs and humans. To this end, I will first review some of the highlights of the last decade or so of canine research. Many of these studies have made their way to the public via news headlines or specials on shows like *60 Minutes*, and like a lot of sensationalized news, the findings are often reported in overly simplified or dramatic ways. Despite how exciting some of the research may be, it will be important to retain the critically anthropomorphic lens when asking questions such as (1) What exactly has been demonstrated? and (2) Is this an isolated finding or one that has been sufficiently replicated and can therefore be generalized among most dogs?

It is my contention that the findings coming from canine research overwhelmingly do demonstrate something special about dogs, although I will not argue that it is as simple as claiming that dogs are smarter than we might have thought. My argument in this chapter is a bit more nuanced and proceeds as follows: First, the research being conducted highlights that although dogs are indeed intelligent, what they are capable of is by no means unique. Many other species have been shown to do many similar things. However, from some of the experiments and what they purportedly demonstrate, it becomes evident that a great deal of what we are learning about canine cognition and affect is complicated by the entangled relationship between humans and dogs. I consequently argue that what we mean by intelligence itself needs to be reevaluated and redefined. For one, despite a long-standing tradition in philosophy of pitting emotion against reason, I argue that studies in canine cognition force us to rethink this dichotomy. Affective intelligence is integral to how dogs interact with humans, and how humans communicate and bond with dogs.

There are other important cognitive skills dogs are repeatedly showing they are capable of, such as *social referencing*. As we will see, the way dogs are able to extract and respond to information in their social umwelt is strikingly similar to the way human toddlers do the same thing. Of course, I will not argue that we ought to think of the two as perfect comparisons, as this would violate the critical anthropomorphic stance I have promised to sustain. The point of comparison I want to focus on is simply the *interactive* component that is fundamental to developing such skills. Likewise, research increasingly supports the claim that nonhuman animals have rich and complex emotional lives,[1] and I argue that as with humans, it is unhelpful to think of the emotions of other animals as entirely internal and individual. This is especially evident when we examine dyadic pairings between humans and domestic canines. The human-dog bond, I suggest, is an unexplored encounter that provides rich and plentiful examples of such coordinated and dynamic processes. At this point in the chapter, therefore, I turn to an argument in philosophy of cognitive science that is overwhelmingly sympathetic to the idea that a proper account of cognition ought to be altered and expanded, and moreover that thinking is not a wholly encapsulated endeavor.

Philosophers have tended to be a collective of skeptics at best, and naysayers at worst, regarding the fruitfulness of studying animal minds. I think this resistance is due in part to an overwhelming loyalty to *cognitivism*, which has been the undisputed frontrunner in theorizing about mentality for quite a while now. A stubborn reliance on a computational account of the mind is one reason that wholehearted allegiance to cognitivism is hindering progress in understanding animal cognition, and specifically in properly assessing the cognitive capabilities of the domestic dog. Turning our attention instead to the *radicals* of philosophy of cognitive science will prove much more promising. Nontraditional, or what I refer to as "radical," philosophy of cognitive science is often summarized as the "4E approach" because it emphasizes the embedded, embodied, enacted, and extended features of cognitive processes (see Newen et al. 2018). Taking this "radical" approach seriously is a large part of the argument I will make regarding how to better approach animal cognition, generally, and to subsequently rid philosophy of its stubborn ties to Cartesianism once and for all. The other key component is to relate this radical approach to the findings in canine science to show that cognitivism fails to provide an adequate account of how dogs think, largely because the cognitive framework with which dogs operate is one of *cooperation* with humans, and is the result of extensive interactions with humans. The same is true of emotions, which I argue, contrary to the typical Cartesian framework, are proper parts of cognitive systems. The idea is that when we look at two tightly interwoven species, such as humans and dogs, the enactive account does well to explain how individuals among those species—through interaction and mutual cooperation—develop collective and shared emotions and cognitive capabilities. This last point, I argue, is a call for further examination and discussion pertaining to coactive cognition and joint affectivity.

Last, as I have indicated already, some of the most fascinating and revealing studies of dog cognition are not solely about the dogs themselves but also concern how they think *with us* and, likewise, how we think with them. Hence, by studying dog cognition, we are necessarily studying our own minds. This *coactive* or *collaborative* cognition cannot be explicated under a cognitivist paradigm and is better captured by the radical philosophers of cognitive science, especially the enactivists. Treating at least some cognitive processes as *cooperative*—as I argue those occurring between us and our dogs must be—is a story about *human* cognition just as much as it is about nonhuman.

Decade of the Dog: An Overview of Recent Findings in Canine Cognition and Affect

Over the last two decades or so, there has been a dramatic surge of interest in the unique abilities of *Canis familiaris*. Before this, it was largely agreed that the family dog was, at best, an obedient, subservient, domesticated *pet*—often capable of some entertaining tricks, to be sure, but not the proper subject of serious scientific inquiry. That dogs were neglected for so long likely stems partly from what we discussed in chapter 1 regarding the rise of ethology and the prescriptions placed on its appropriate study. Once Tinbergen and Lorenz convincingly established that animals observed in their natural habitats, free from the human tampering associated with laboratory settings, behaved quite differently, it became crucial for ethologists to ensure genuine ethological methodology. Dogs had been studied in the past, most famously by the pioneer of classical conditioning, Ivan Pavlov, whose salivating dogs paved the way for Skinnerian radical behaviorism. However, this sort of contrived experimental setup is generally frowned upon by ethologists, as behaviors that are the result of operant conditioning are not "natural" and would most likely not occur if the animal were left to its own devices in its typical habitat. Dogs are inextricably bound to "unnatural" surroundings and, worse, are constantly intermingling with humans; hence, they received very little attention once ethology took hold in the animal science world.

As I argued in the first chapter, however, determining the "natural" habitat of a dog proves to be difficult if not impossible, unless of course we allow that the denotation of the term "natural" as it applies to dogs might indeed include *with humans*. After all, there is no other place dogs live. Even feral or "street" dogs interact with the human socio-technological lifeworld, so studying dogs free from human influence not only is impossible but makes no sense if we are taking the ethological dictum to "only study animals in their typical surroundings" seriously. In sum, with dogs, ethologists face a dilemma: either assume no good scientific research can come from studying them, because of their entanglement with humans, or accept that human entanglement is *part of their umwelt* and attempt to study them anyway.

Recently, many researchers have aligned themselves on the second horn of this dilemma, devising all sorts of methods to assess how and why dogs behave the ways they do. Brian Hare, for instance, has performed countless studies indicating canine intelligence, including social referencing skills and abstract reasoning abilities.[2] At the Family Dog Lab,[3] researchers have demonstrated similar capacities through experiments geared to test the connectedness between humans and dogs. Of course, the sorts of experiments performed in laboratory settings are likely to spark controversy among ethologists. Thus many of the studies, especially those that form part of Hare's Dognition games, are set up so they can be performed as naturally as possible, in the homes of dogs and their human companions. The way Dognition works is that families set up accounts online, follow the instructions to test their dogs on a number of dimensions, and then record the results on the website. The data are saved at Hare's lab, and this information helps him and his team assess how dogs all over the world respond to certain scenarios. One of the most revealing findings for Hare—indeed, one reason he shifted his focus from great apes to dogs—was the discovery that dogs will follow human pointing to find food or toys. With a human hiding food in one of two cups and not allowing the dog to know which cup the food is hidden in, then pointing to a cup, the dog is overwhelmingly likely to head toward the cup being pointed to, even if there is no food there. This is one of the tests people can run in their own homes with their own dogs. And it turns out that nearly all dogs are quite adept at this, with no priming or training. For Hare this finding was important because our closest genetic relatives—chimpanzees and bonobos—do not pay attention to our social cues, such as pointing or gaze, at least not with the frequency and reliability with which dogs do, and not without training. From an ethological perspective, it is arguable that dogs utilize this skill in order to better adapt to their environments. In the case of following a human's pointing, this skill might have arisen as a result of the need to cooperate with humans, which itself was beneficial to the species and individuals among the species, since cooperation with humans often results in food, shelter, and other forms of care conducive to survival and proliferation.

Perhaps findings such as those Hare has gathered from the Dognition website would not count as the kind of intelligence Descartes had in mind when he discounted the possibility of "automata" having minds, and hence another reason dogs might have been ignored for so many years. Consider this quote from his *Discourse*:

> For one could easily conceive of a machine that is made in such a way that it utters words, and even that it would utter some words in response to phys-

ical actions that cause a change in its organs.... But it could not arrange words in different ways to reply to the meaning of everything that is said in its presence, as even the most unintelligent human beings can do.... Thus one would discover that they did not act on the basis of knowledge, but merely as a result of the disposition of their organs. For whereas reason is a universal instrument that can be used in all kinds of situations, these organs need a specific disposition for every particular action. ([1637] 1999, 40)

The Cartesian "mark" of cognition—that is, rationality—has been argued to only occur in animals that share neuroanatomical features with or are genetically close to humans. Dolphins, with their large cerebral cortexes, and apes, our closest relatives, are often focused on because they have the highest chance of possessing this "universal instrument" of reason.[4] Chimp signing, for example, would likely give Descartes at least some pause for reconsideration of his meat machine manifesto.[5] But even more disturbing for Descartes might be to learn that there are in fact examples of dogs who are arguably just as adept with the meanings of words as chimps, dolphins, and three-year-old toddlers. The famous border collie Chaser, who seemingly knew the meanings of more than a thousand words and could even discern the difference between the way verbs and nouns function in longer sentences, is one of many cases in which researchers are demonstrating abilities in dogs that were once thought achievable only by humans and perhaps apes.[6]

Chaser the border collie made headlines as "the world's smartest dog" when her guardian, John Pilley, and his colleague, Alliston Reid, published a 2011 article in *Behavioural Processes* claiming that Chaser comprehended names of objects as verbal referents. Not only was Chaser shown to be able to map proper nouns onto individual objects, but also to discern between common nouns, such as "ball" or "disc" and proper nouns, and to distinguish among verbs and nouns, and to connect them in commands given to her. So, for example, when asked to fetch "a ball," she would get any object that counted as a ball, but when asked to fetch "Inky," she would get the toy specifically named Inky. If asked to take Inky to Sugar, she would pick up the toy named Inky and take it over to the toy named Sugar. She would also place her foot on a toy when told to "paw" it, as opposed to fetching it or "nosing" it.

Chaser was not the first dog to exhibit the capacity for "fast mapping" (cf. Kaminski et al. 2004), but she was instrumental in bringing these findings regarding the capacities of canines into broader public view. After Pilley and Reid published their findings, Chaser was featured on *Nova Science Now* (Cort 2011) and *60 Minutes* (Cooper 2015) and appeared in several other news and science publications. But Chaser was just one dog—and a dog who was extensively trained by several researchers. While her behaviors might rival those of two- or three-year-old children, it would be far too presumptuous to argue that the mechanisms underlying those behaviors are the same. Pilley and Reid acknowledged this and suggested more research was needed to determine to what extent all dogs are capable of similar skills, and what subtends those skills.

As mentioned above, a finding that has been particularly intriguing—especially since chimpanzees and bonobos, our closest genetic relatives, pale in comparison to dogs on these tests—is the ability dogs seem to possess when it comes to social referencing. In experiments, they have been shown to follow the pointing gestures of humans, as well as humans' gaze (Hare et al. 1998, 2002; Hare and Woods 2013). Brian Hare and his associates have even demonstrated that dogs will use human feet as indicators of intention. When food is hidden under one of two containers, if a human uses her foot to point in the direction of one of them, dogs overwhelmingly go toward that container, even if no food is under it. The capacity for reading intention through gestures such as these is crucial for human infants and toddlers as they embark on learning language and adapting to human culture. The fact that dogs have repeatedly shown themselves capable of the same skill, while other animals are much less consistent with this ability or lack it entirely, suggests, at the very least, that the long-standing relationship between humans and dogs is partly responsible. From an evolutionary standpoint, one might also speculate that it was to the dogs' advantage to learn how to read human intention, so that cooperation and ultimately food rewards could follow. Whether the social referencing capacities of dogs implies that they can engage in full-fledged *mindreading*—the ability to discern what others are thinking by observing their behaviors—remains an open debate. One could easily argue that these studies just demonstrate an associative learning process that has taken place over many centuries. We revisit this debate throughout the book. For now, it is rather unanimously agreed that dogs are quite good at reading human gestural cues, most likely better than any other species. Thus we have described a unique *behavior* of dogs. Providing an *explanation* of that behavior will require more discussion and thought.

Related to the social referencing capacities of dogs, another finding that has captured the attention of scientists and philosophers pertains to the way dogs perceive human faces. Humans have a left-gaze bias when assessing the face of another human. Because more of the emotional information in our facial expressions is found on the right side of our faces, in a preconscious and automatic instant we scan from left to right—that is, beginning at the opposite person's right side. The theory is that we do this to quickly grasp the affective state of the person with whom we are interacting. Using eye-tracking software, Racca et al. (2012) have shown that dogs will utilize the same left-gaze bias when looking at human faces. For control purposes, the experimenters also had dogs look at other dog faces and non-faces, and found that the left-gaze bias was present only when looking at human faces. Again, this is just a behavior that has been discovered, and it can be intoxicatingly easy to jump to the conclusion that dogs have evolved a special facial recognition module identical to the one humans have, or that they are capable of reading emotional information from our faces better than any other animal can.[7] Nevertheless, as I have repeatedly urged, a critical anthropomorphism demands that we exercise caution when attempting to explain behaviors, even if it is fairly obvious that those behaviors are analogous to human behaviors. The questions we must ask about this finding are even more complex than those regarding social referencing. For one, it might be the case that dogs evolved this ability simply for survival. An angry human face intent on killing a wolf to eat it is vastly different from a timidly friendly face beseeching cooperation. Learning those differences might have been key to the domestication process. All of this could be explained without reference to a canine mental process of *understanding* emotions. Furthermore, the study carried out by Racca et al. has not been replicated sufficiently, and it remains to be seen if dogs in general possess this left-gaze bias, or if dogs from different parts of the world, or dogs living less proximally to humans, such as street dogs in India, might not have this skill.

Another finding worth noting comes not from studying dogs' behavior, but from their physiology. When humans interact with their dogs, their brains release the hormone *oxytocin*. This substance is often referred to as the "bonding hormone," as it is thought to facilitate bonding between humans. Oxytocin floods the brains of female humans just after they have given birth and when they are breastfeeding their babies. Researchers have found that a similar oxytocin "bump" occurs when humans pet their beloved canine companions. Even more astounding is the fact that oxytocin is present in dogs' brains when they are being petted and even when they are simply looking at us (Hare and Woods 2013). Brian Hare has referred to this as dogs "hugging you with their eyes" (Cooper 2015). As you can see, even someone so committed to serious scientific inquiry cannot help himself when it comes to bouts of unwarranted anthropomorphism. While it might be overblown to claim that dogs are hugging us with their eyes simply because they have a flood of oxytocin, again, this is compelling evidence that the interactions between humans and dogs are at least partly responsible, not just for behavioral modifications, but for biological ones as well. In attempting to explain to what extent dogs care about or love humans the way humans love and care for them, again, caution must be exercised. Oxytocin, for example, is not always the rosy, happy, and loving hormone it is often touted to be. In some cases it can be associated with aggressive and violent behavior, as well as PTSD (see De Dreu et al. 2011; Herzog 2011; Guzmán et al. 2013). So when assessing the similitude between humans and dogs regarding the expression of this hormone, we must also keep in mind that it's not entirely agreed what the function of oxytocin is *for us*, let alone for our dogs.

Last, there is compelling evidence regarding the capacity for love and other human emotions in dogs that comes from neuroscience. Researchers have been able to perform functional magnetic resonance imaging (FMRI) on dogs, which alone is a feat. The dogs have to be trained to sit still in the machine, which is noisy and likely scary. Given protective ear coverings, however, dogs seem happy to sit for the test. This is significant because before such modifications and training, the only brain scans that could be done on dogs were accomplished while the dog was sleeping. Dogs fully awake are placed in FMRIs and are presented with objects, such as cotton swabs, with the sweat of a human on it. Unsurprisingly, the part of their brain that processes olfactory input is activated. When they smell the sweat of their guardian, however, not only is the olfactory center activated, but so is the caudate nucleus, the part of the brain that recognizes reward—the same part of a human's brain that is activated when we anticipate being reunited with a loved one, or when we are engaging in our favorite activities (Berns 2013). This is further reason to take note of dogs and how they are showing more and more that they possess some unique behavioral and biological traits heretofore overlooked. What all these findings imply regarding what dogs are actually *thinking* or *feeling* is a question that deserves careful consideration before making an argument.

At this point, however, rather than delving into a discussion about how best to explain all these scientific findings, I want to take a detour and, as promised in the introduction, examine another revolution in thinking about minds that has been taking place recently. Much as cognitive ethology has expanded its purview to include dogs, philosophy of cognitive science has been undergoing a sort of "extending," with the meaning of the term "cognitive" a matter of much debate. Many thinkers, myself included, have argued that what we mean by *cognitive process* should not be limited to an event taking place inside a brain, human or otherwise. Instead, thinking is most often a hybridized set of neurological, physiological, technological, and even sociological processes. This movement away from standard cognitive science and toward more radical approaches is paving the way for many advances in the field, chief among which I think is a successful approach to studying animal cognition. So in the next section I outline this *radical* approach, how it is being utilized to reconceptualize human cognition and affect, and how it can be applied to the study of canine cognition. All of this will, I hope, provide a grounding from which to tackle the questions surrounding a proper *explanation* for all these recent findings.

Standard versus Radical Philosophy of Cognitive Science

Philosophy of cognitive science has in its purview a vast range of topics, fields, and disciplines, such as intentionality, nonhuman animal thinking, neuroscience, psychology, and linguistics. So it stands to reason that a unifying account of what practices are acceptable within this study would be difficult to provide. Nonetheless, what we might call traditional or standard cognitive science has been dominated for some time now by the view that the mind is analogous to a computer. For example, the physical symbol systems (PSS) approach (Newell and Simon 1976) maintains that the brain operates by receiving data in the form of symbolic input, manipulates these data in the form of internal representations, and then produces an appropriate output. Unsurprisingly, since this processing mimics the way a computer operates, PSS guides a large amount of research in artificial intelligence.

The umbrella term under which these computational strategies can roughly be placed is *cognitivism*. Albeit different in important ways, the various strands of cognitivism are united by the idea that cognition is a rational process. Additionally, in a cognitivist paradigm, the symbolic and linguistic components of thinking are emphasized, and some go so far as to say that *all* cognition is representational. Further still, most cognitivists argue that cognition is a process that occurs within the confines of a closed physical system, the most obvious choice for such a system being the human brain.

This last point would make it seem that cognitivism is a suitable rejection of the Cartesian ideology mentioned at the beginning of this chapter. After all, the idea that the mind is a physical system is certainly not what Descartes envisaged, and furthermore, cognitivism is more likely to grant nonhumans, such as computers, some form of "rationality," while Descartes was very clear only humans possess this "universal instrument." However, despite the metaphysical rejection of substance dualism, computational approaches to cognition share more in common with Descartes than a first glance might reveal. Recall, for instance, that the functionalist approach is not concerned so much with the *type* of system in which cognition inheres, but rather with the *processing* itself. It could be that a mental state such as fear is realized by a brain, a computer, or a robotic elephant. So long as the "total states" of each system are functionally equivalent—that is, the inputs, internal computations, and outputs all function the same in the overall system—then we ought not to hesitate in claiming that what we are observing is fear in all three cases. When you consider cognition in this way, it begins to seem more like an abstract idea, a process that is not beholden to any one type of physical system. If the right sorts of processes align, thought will occur. Again, this is not Cartesian dualism, not by a long shot, but under a liberally functionalist approach it is not inconceivable to think of cognition as something wholly distinct from the brain. More important, cognitivism shares with Cartesianism the assumption that thinking is, by definition, a rational process. Indeed, rationality is the universal instrument of thought. How else do we go about making decisions, solving problems, and having meaningful conversations?

Another way cognitivism can seem appealing is in explaining affect. Philosophers have grappled with how best to classify emotions for centuries, so I will not delve into a comprehensive treatment here, but it is worth spending a little time looking at how emotions are typically understood in the philosophical literature, how cognitivism is often utilized to explain them, and to what extent these explanations cohere with biological accounts of affect.

The "standard" approach to studying emotions, either within philosophy or among other disciplines, generally encompasses those theories or frameworks that regard emotions as private, internal, intracranial phenomena. To be sure, if we survey the major contenders within this paradigm—neo-Jamesian theories, intentional theory, appraisal theory, and perceptual theory, among others—important distinctions exist, and to claim that all of these "standard" accounts treat the ontology of emotions in precisely the same way would be an overstatement. While intentional theorists argue that emotions are always *about* something, a more cognitivist approach or an *appraisal* theorist would argue that emotions are fundamentally judgments. Despite these differences in classification, I am grouping these theories together based not on *what* emotions are but instead *where* they are.

That is to say, according to most standard accounts, emotions take place in a sort of Cartesian theater, within the confines of the brain-body, inaccessible to others except via behavioral analysis, and, most important, are individual and subjective.[8] Hence, cognitivism seems well equipped to explain affect.

Though it might be plausible that phenomena such as sadness, fear, or joy are best considered subjective, a careful taxonomy of emotions reveals more than just these sorts of private and temporary feelings. For instance, we can distinguish between emotions that are *primary* (immediate feelings like anger or happiness) and *secondary* (those that result from a conglomeration of other feelings, such as guilt or shame). Secondary emotions are not only more complex than primary emotions, but they are often social in nature and thus not entirely private. Shame, for example, implicates the other insofar as when experiencing this emotion, I am not just sad for myself but am remorseful for what I have done in the eyes of someone else. Thus it is more appropriate to think of these secondary social emotions as intersubjective—constituted in the interaction between two or more individuals.

Intersubjectivity has received more attention recently in the sciences, particularly as it relates to emotional development. The idea that interactions are primary for and constitutive of emotional awareness and regulation has its roots in infant-caregiver studies. The "still face experiments" of Tronick et al. (1978) and more recently Adamson and Frick (2003) demonstrated that babies are already primed for interaction at an early age and are distressed when it is unavailable. When parents purposely ceased interacting with their babies and presented them instead with a "still" or blank face, after a few moments the babies became agitated, began making gestures to try to elicit interaction, and eventually were so stressed that many of them cried or screamed. Meltzoff and Moore (1977) famously showed that neonates as young as ten minutes old will imitate facial expressions of caregivers. While this ability is arguably not a conscious or purposeful act, it is also not merely a reflex, because infants won't mimic or pay attention to just anything. It is specifically human faces that elicit interaction. As the infant grows and the bond with the caregiver strengthens, this ability to read facial expressions, gestures, and body language sharpens to the point that prelinguistic infants are able to determine whether to traverse a seemingly dangerous "visual cliff," all based on the emotional cues of the caregiver (Gibson and Walk 1960; Campos et al. 1978; Sorce et al. 1985).

Studies such as these have led some philosophers (Trevarthen 1979; Gallagher 2001; Daly 2014) to argue that intersubjectivity is primary and the idea of a private, individual self or subject comes later. In other words, the ability to read others' intentions and to understand what they are thinking and feeling comes about by interaction, and this interaction happens prior to any theorizing, introspecting, or analogizing. Most if not all early interactions are nonlinguistic (they may be verbal, but the infant is not recognizing them as such and is not responding in kind) and are heavily based in emotional cues such as smiles, high tones of voice, and bodily gestures indicating feeling. It stands to reason, therefore, that our ability to experience and comprehend complex social emotions, such as shame, guilt, or pride, has its roots in these early interactions where we learn to distinguish our own thoughts and emotions from those of the other. The idea that interaction is fundamental to emotions gains traction when we consider studies of infant-caregiver co-regulation and the development of the autonomic nervous system, which is intimately connected to the limbic system. Relatively nascent theories, such as the polyvagal account of co-regulation, have been taken more seriously in recent years, especially in psychotherapy, where techniques that depend on mutually interdependent regulating of the patient's vagus nerve and autonomic nervous system have been shown to be effective treatments for PTSD, anxiety, and depression (Porges 2011). In short, much of what humans *feel* is entirely dependent on how they *interact* in a social world.

Nothing of what I have said so far directly implies that social emotions are not experienced as private and subjective affairs. However, some thinkers (e.g., Fuchs 2002) have interpreted these findings, coupled with phenomenological descriptions of social emotions, to indicate that emotions such as shame or pride are better thought of as intersubjectively constituted. Some go so far as to claim that a lot of the emotions that are social in nature are often collective and shared (see Scheve and Salmela 2014). Think, for example, of the sense of pride and excitement you might feel at a basketball game. Again, one could appeal to subjective phenomenology here—it is *I* who am feeling *my* specific pride about my team's recent score—but it is arguably the case that at least some aspect of this pride is shared and distributed among all the other fervent fans sitting near you. As you cheer, they cheer, and as they get louder, you get louder. A "wave" might even form as arms rise and fall in a ripple effect around the arena. It is as if the emotions are contagious at this point. And much like other contagions, they are spread easily through interactions. Pride, at least in this instance, is collective.

Last, if we consider a third sort of emotions—*moods*, or what I will henceforth refer to as *background emotions*—the case against an overwhelmingly subjective account of affect is made all the stronger. Background emotions are the least frequently discussed in the philosophical discourse surrounding sentiment. A background emotion, unlike other feelings that come and go, is longer lasting and often affects multiple aspects of a person's life. Depression, for example, is not just sadness but is deep and temporally extended sorrow and suffering. Depressed persons are often unable to enjoy activities they might otherwise take pleasure in, and even their physiological comportment is often "depressed" in the sense that basic movements such as getting out of bed or eating are impaired.[9] To be sure, one can find copious mention of depression and other sorts of long-term emotions within philosophy of psychology, but the discussion of these phenomena is typically more about how the disorder functions, how it is diagnosed, and how it can be treated. Less time has been devoted to investigating the phenomenology of background emotions and how they differ from other emotions in the taxonomy we have so far discussed.

In sum, as compelling as cognitivism can be, it is not without its detractors. Most notably, the claim that cognition is a rational, symbolic, and representational process has come under scrutiny in the last several decades. Similarly, the role of the body and its interaction with an environment has been shown to be much more important than cognitivists would have us believe. And emotions are not simply individual feelings, according to many accounts of affective cognition. It is my contention that all these shifts in focus align well with the ethological revolution insofar as the situatedness and ecological aspects of cognition and affect are elevated to a much greater place of importance. Let us, then, consider some of these "radicals" in philosophy of cognition and emotion.

Numerous sources could be cited as exemplars of this newer, more radical cognitive science. Arguably, however, many of the current trends have their roots in a highly influential 1991 book, *The Embodied Mind*, in which Francisco Varela, Evan Thompson, and Eleanor Rosch levy a convincing and potentially devastating attack on cognitivism. The conclusion, they argue, is that cognition is an *embodied* practice, not merely an *embrained* one. Taking cues from this radically embodied thesis, thinkers have gone on to push even more contentious views, such as the argument that thinking is not even constrained to the physical body.

Andy Clark and David Chalmers (1998) have claimed that cognition is often distributed among human organisms and the tools they use. If I am using my GPS to find my way to a restaurant, the beliefs I form about the restaurant's location are not entirely in my head, but are rather spread among my brain and the technology to which I am coupled. Similarly, those who stress the enacted nature of cognition focus on the dynamic coupling of human organism and its environment. Drawing from the work of James J. Gibson (1977, 1979), for instance, enactivists such as Evan Thompson (2007) and Hutto and Myin (2012) claim that the world affords us potentials to act and that it is in this interplay between brains, bodies, and environment that thought emerges. There are important differences between the enacted and extended views, but they are not relevant to my overall argument. What ties these various pictures of cognition together is that they all describe cognition in terms of rich embodiment and coupling with the environment and argue that thoughtful action occurs in a myriad of ways overlooked by traditional cognitive science.

Similarly, many of the so-called radicals in philosophy of cognitive science tend to view emotions as ecological, or even extended beyond the confines of the biological body. Recall that the third type of emotions—background emotions, or moods—can emerge and be sustained in interactive exchanges. Ratcliffe (2010), for example, describes such emotions in terms of spaces of possible actions. Background emotions, he argues, form the relations we have with the social world. In a similar vein, Fonagy and Target provide a detailed phenomenology of depression wherein they argue that it is often experienced as an infantile loss of proximity and connectedness to others, such that "the phenomenology of depression may not be readily understandable without considering the involvement in it of an experience of lack of shared consciousness (outside, as well as inside)" (2007, 934). This makes depression seem much more like a background against which we experience alienation and loss of connectivity to the world, often resulting in an inability to function in otherwise normal ways. Indeed, depressed persons often complain of feeling heavy, immobile, and unmotivated. Likewise, they often withdraw from the social world, despite needing proximity to others to feel better. Another key experiential element of depression is that, like other background emotions, it seems permanent, rigid, and inflexible to the patient, and thus impossible to escape. As Karp (1996) notes, much of the pain arises out of the recognition that what might make the depressive feel better—human connection—seems impossible in the midst of a paralyzing episode of depression.

Depression and other background emotions, along with social emotions such as shame, pride, and guilt, demand a different phenomenological description from that of primary emotions, to be sure. But some theorists go further, and argue that this phenomenological difference undergirds an even more fundamental difference. Jan Slaby argues that some emotions are constituted by *phenomenal coupling*—that is, "the direct, online engagement of an agent's affectivity with an environmental structure or process that itself manifests affect-like, expressive qualities" (2014, 41). This coupling works by interaction within an "affective atmosphere." Take collective grief as an example. It is often difficult to resist feeling sad alongside those who are grieving, and certain environments exacerbate this contagious effect. The affective atmosphere is a dynamic backdrop against which the social interactions take place. These backdrops are functionally similar, Slaby claims, to the tools and technologies we often couple ourselves with in other forms of cognitive enhancement, an argument that echoes several varieties of the extended and enactive accounts of cognition.[10] Thus emotions, like other forms of cognition, can be extended beyond the brain and distributed among lived bodies and environments.

These recent trends in cognitive science represent what I have been referring to as radical stances. They are radical only insofar as they break with the standard, commonly accepted framework for explaining cognition. As I have argued elsewhere (2013; 2015b), the radical arguments are actually not that radical at all—instead, they are the most accurate and explanatorily powerful account of the dynamic and distributed collection of processes we call "the mind." The debate between the "standards" and the "radicals" is long-standing and complex, and it won't be settled in this book. That is not my aim. In suggesting that the radicals might have it right, I am merely asking us to think about what would result if these positions were taken more seriously in cognitive science, and especially with regard to nonhuman animal cognition.

The version of radical philosophy of cognitive science I want to examine is the enactivist position, and how it might pertain to cognitive ethology generally and canine cognition more specifically. Enactivism makes the point most often that cognition is likely occurring in all sorts of different ways across the animal kingdom. Many enactivists insist that other forms of life are readily capable of intelligent problem-solving behavior that ought to count as "cognitive" even if those behaviors don't smoothly translate into human ones. I will assuredly never know what it's like to be a bat, but I can appreciate how hearing, instead of sight, might guide my movement. And it is similar for other animals; I shall surely never *be* a dog and thus can never fully understand the complexity of its olfactory system, but I can recognize that the olfactory system is a rich source of information for the dog, as rich, probably, as my visual system is for me. The dog's cognitive niche is shaped by what it smells. In other words, the dog's nose is part of its cognitive machinery.

Another way enactivism might approach nonhuman animal minds in a critically anthropomorphic way is by adopting a strong sensorimotor model (SSM) of perception (O'Regan and Noë 2001; Noë 2004). Not all enactivists endorse SSM, but for those who do, such as Alva Noë, the account goes a step beyond claiming that perception and action are tightly intermingled, and argues that action *constitutes* perception. In perception, according to SSM, we draw upon implicit knowledge of how our past actions have influenced the various ways in which we sense the world. These *sensorimotor contingencies*, as they are dubbed, are relations obtaining between movement, bodily comportment, and sensory stimulation. They in turn impact how we engage the world at any given moment. The act of seeing, Noë argues, is more like touching, in that to look at an object is to engage with it spatially. Even if you cannot see the entire object, your visual system "reaches around" to the other sides that are not present. It fills in what is missing, based on how your body stands in relation to the object, as well has how your prior engagements with objects like this one have shaped your visual experiences. This is a radical view of cognition because it amounts to claiming that perception *just is* a bodily *know-how*, which further means that perception will be highly dependent on one's prior sensorimotor dependencies. In fact, this is one of the criticisms of SSM: that it amounts to a chauvinism about perceptual abilities. In order to have perception *like ours*, for instance, a creature must have a body *like ours*.

An effective rejoinder to this objection is to simply agree with it. Indeed, walking around on four legs as opposed to two should make some important differences in how a dog perceives the world compared with a bipedal human. As we just discussed, a dog's olfactory system is uniquely suited to making sense of the world through smell, and so it stands to reason that dogs have different phenomenological experiences of the olfactory world than humans. If the worry is, however, that a system becomes so hypersensitive to the differences that obtain at the finest levels of experience—a tiny difference in the way the olfactory system is constructed might result in the most minute difference in the way that smells are processed—then this should concern us, not just about drawing parallels between human and nonhuman perception, but also about making human-to-human comparisons. Surely my sensorimotor dependencies have differed greatly from my best friends' over the years, and those differences have likely shaped how visual, tactile, auditory, olfactory, and gustatory sensations are experienced, processed, and catalogued in each of us. The real pressing question for proponents of SSM is, it seems, *To what extent do those differences make a difference?* This puts us right back at one of the open empirical questions in cognitive ethology and philosophy of cognitive science, namely, *How can we understand what another creature is thinking or feeling if we are not able to access their subjective minds?* In other words, this supposed objection to SSM really just highlights a question that needs more empirical work to properly answer it. Moreover, if we adopt SSM, we might be in a better position to frame the question itself, given that proponents of SSM deny that perception is something hidden away *inside* a creature. As Noë argues, "the enactive view denies that we represent spatial properties in perception by correlating them with kinds of sensation. There is no *sensation* of roundness or distance, whether tactile, visual or otherwise. When we experience something as a cube in perception, we do so because we recognize that its appearance varies (or would vary) as a result of movement, that it exhibits a specific sensorimotor profile" (2004, 101–2). Put differently, if we wish to know how some creature is experiencing the world, we need not look to its inner mental representations; we need to see what it is *doing*, how it is acting out perception-in-its-activities, and how it responds to the varying sensorimotor patterns enacted by those movements.

Adopting SSM would arguably mean that enactivists can easily allow for nonhuman thinking to exist while also carefully navigating the anthropocentrism issue. We cannot avoid anthropomorphizing entirely. There is no way to think as a nonhuman. However, we can—indeed we *must*, claims the enactivist—begin to recognize other ways of thinking that might not look anything like what we are accustomed to as humans. This widening of focus also applies to our own cognition. Emotions and social dynamics have long been ignored in the philosophy of mind. But as I will discuss in detail in the next chapter, more attention is being paid to the ways in which feelings and embodied interactivity shape thought, and this increased attention most often comes from the radicals, especially the enactivists. Their interest in affect and social cognition, along with the general enactivist claim that many life forms besides humans think, represents some of the most emphatic movement away from Cartesianism thus far.

I mention affect and social cognition precisely because these forms of cognition are *not* examples of the supposed rational and linguistic faculties Descartes and other cognitivists have in mind when they talk about "thought." Instead, these modes of thinking indicate a different sort of intelligence from the type Descartes so venerated. Nevertheless, when a dog demonstrates a capacity for empathy, love, or mindreading, I am arguing that this ought to count as intelligence all the same. Much as with humans, the role of emotion and social referencing should not be ignored. Indeed, it is this ability to read facial expressions and follow gaze and pointing that provides the foundation for language and culture in the infant human. Thus, as Chaser's guardian John Pilley has argued, we ought to start thinking of our dogs as toddlers rather than as pets.[11] While I tend to agree with Pilley here, especially considering my own interactions with the various dogs and toddlers in my life, I do want to caution against accepting his analogy entirely. There are similarities between dogs and babies, to be sure—much as I cannot quite comprehend what is going through the mind of a preverbal child as she plays in a world of her own imaginative creation, I also don't fully grasp what is going on when my dog Tesla stares for hours on end at the cat. A key difference remains, however: the preverbal toddler is assuredly on her way to full-blown adult linguistic aptitude, while my dog is not. Thus, to claim a dog and a toddler are entirely on par with each other is (1) an oversimplification and (2) an example of egregious or uncritical anthropomorphism.[12]

But there is another component of cognition where the parallels between humans and dogs are perhaps the strongest: affect. To be sure, dogs show emotions in different ways than humans, but those differences are not sufficient to make affective communication between the species impossible. In fact, the recent findings we have so far discussed suggest that in affective interaction, humans and dogs appear best equipped to resonate with and understand one another. This is arguably because emotions are subtended by many of the same mechanisms in both species.

Taking these recent findings regarding canine cognition, along with the general shift in cognitive science to a more ecological approach to intelligence, if we then consider emotions in light of these trends, we can tell a similar story. Dogs, like humans, have complex emotional lives, and experiments bear this out. In addition, it turns out that the emotions dogs experience are not mere internal feelings or purposeless phenomena. As with other cognitive adaptations, dogs have evolved to love, grieve, fear, trust, get depressed, and feel lonely, and the most current and compelling research suggests they have developed these emotional capacities to better live alongside us. Likewise, humans have learned to become more attuned to dogs after thousands of years of thinking *and* feeling with them.

An example that illustrates how evolution likely favored those wolves who were emotionally attuned to humans comes from research on another member of the *Canidae* family: *Vulpus vulpus*, or the silver fox. These foxes, who are commonly referred to as "red foxes," are more typically silver-black when found in northern Siberia, where this research was originally conducted (see Trut 1999; Trut et al. 2004, 2009). Dmitry Belyayev began breeding the foxes in the 1950s, operating under the hypothesis that behavioral traits could influence physiological ones. His suspicion was specifically that "tameness" would, over time, alter the way the foxes appeared. "Tame" is a tricky word—it tends to connote docile, trained, and domesticated. And to some extent, this is what was being bred into the foxes. Only the ones who did not attack Belyayev and his research associate, Lyudmila Trut, during feeding, were permitted to reproduce. But on closer reading, the behavioral traits of those foxes allowed to reproduce could arguably be described as "trust," "kindness," or "cooperation." These are loaded terms themselves and could be seen as uncritically anthropomorphic, but it seems that any word we choose to describe the interrelationship between the humans and those foxes permitted to reproduce will involve some degree of creative license. Indeed, they are human terms and, in that sense, are *anthropocentric*—"tame" being perhaps the most highly anthropocentric of them all. What matters here is that over generations, as more and more foxes who were tolerant of humans bred more of the same types of foxes, each subsequent generation changed drastically in appearance. Ears became floppier, tails bushier, and even the fur color changed to include more whiteness and spots. The more the foxes showed what we might call *pro-social* interactivity with humans, the more they began to look like puppies—the infantile *Canis familiaris* with whom so many humans are so incredibly enamored. In turn, it became easier for the researchers to trust in and engage with the foxes, as they knew on the basis of appearance alone which foxes were most likely not to harm them. The relationships forged between the researchers and the foxes began to look much more like the human-dog bonds I've been describing in this book.

The Siberian fox experiments are problematic, to say the least, for various reasons, perhaps chief among them being concerns over animal welfare. Luckily from that angle, the experiments have ceased, but they are still informative and thought provoking in terms of genetics, epigenetics, and the interrelationship between behavioral traits and physiology. For our purposes here, I discuss the experiments to demonstrate that it was precisely the *interactions themselves*—the process by which the humans and foxes became affectively co-attuned—that drove these changes along. As the foxes learned to trust the humans, and in turn the humans learned which foxes would afford positive interactions, the two species became more emotionally invested in the interactions. It's an open empirical question whether the foxes genuinely felt love or any other positive emotion for the humans, but from a purely behavioral standpoint, it is clear that they were positively influenced by the pro-social engagements and that those transactions served to reinforce future positive exchanges. This is arguably how it happened as wolves self-domesticated, albeit in a much less contrived and plausibly less inhumane manner. And as more generations of wolves-with-humans produced affective dyads that were positive and fulfilling, the fluffier, smaller, more spotted and floppy-eared dogs so many humans call "best friends" slowly emerged (Hare et al. 2012).

Let us return to the dog-human research we have discussed and consider one of Hare's tests—"to sneak or not sneak"—wherein dogs are told to leave a treat and their owners turn around or cover their eyes. Many dogs will sneak up and grab the treat when they know they aren't being watched, but some will not. One interpretation of these results is that the dogs who sneak are more intelligent and "know" that the humans are not watching, while the dogs who do not sneak are blindly obedient and perhaps rather dull. A different interpretation, however, is that the dogs who sneak are more cunning than the dogs who do not, while the dogs who obey the command, even when their owners are not looking, are more attached and attuned to their owners, and perhaps more trusting. In assessing the varieties of intelligence dogs possess, Hare's work shows that some dogs will score much higher when it comes to cunning, but this does not make them "smarter" than other dogs. Being a trickster is just one way of adapting to an environment, while another equally effective way, one that many of the more emotionally connected dogs have exploited, is delaying gratification and trusting in the human counterpart to deliver the goods as long as the desired behavior is demonstrated. My dog Darwin, for example, is not a trickster, but when tested on his ability to determine where treats are located, rather than using his nose, he follows my point, my gaze, and even my foot pointing 100 percent of the time. Dogs are clever in all sorts of ways, but one method by which they have cooperated with humans—by becoming emotionally proximal to us—has just recently been uncovered and explored, thanks in part to Hare's research.

I have chosen to focus on dogs, not just because of their unique social referencing and emotional capacities, but also because I think the minds of dogs have shaped our own human minds in unique and ultimately positive ways. One of these fruitful alterations comes in the form of reconceiving the very definition of thought. In other words, I am arguing that examining the way dogs think—and, more important, how they think *with us*—forces us to change some of our fundamentally held ideologies about the human mind and how it works. Not only does the interactive process between dogs and humans highlight the ways in which thinking is a lot more social and emotional than we might ever have imagined, but dogs have also been shaping us as a species. Contrary to long-held assumptions that humans domesticated dogs in a purposeful and targeted way, the new and more convincing evolutionary theory is that wolves self-domesticated because of scarcity of food and, strangely, competition with the only other alpha hunter around: *us* (Coppinger and Coppinger 2001). As Shipman (2015) has argued, rather than compete with us, wolves essentially *used* us, scavenging food we left behind while also experimenting with how close they could get to us. In a careful dance, humans and wolves engaged in a mutual building of trust over the years. Our very survival as a species, we might say, depended on our canine friends, and theirs was highly dependent on ours.[13] Taking these points about the collaboration between humans and dogs seriously, in the next section I shall discuss one of the central tenets of the book, the idea of "coactive cognition," and how this enactivist-based dual-partner type of thinking further dismantles standard cognitive science and its lingering attachment to Cartesianism. But it will be important to keep in mind that cognition is not always rational and linguistic, nor is it necessarily computational in nature, and the emotional lives of both humans and dogs are just as much a part of the proper study of their "minds," and hence their "co-affectivity."

Coactive and Co-affective: Thinking-with Dogs

I have argued that "dog smarts" force us to expand the limits of what we mean when we use the term "cognitive," because the way dogs think does not fit neatly within a cognitivist framework. Dogs think in a uniquely canine way, and I have interpreted the studies regarding their cognitive abilities in a way that stresses the distinction between "us" and "them." Nevertheless, there is compelling reason to believe a lot of our experiences overlap phenomenologically. In other words, it is not an egregious and uncritical anthropomorphism if, upon finding that dogs' neurobiological processes when interacting with us are similar to ours when interacting with them, we conclude that humans and dogs both enjoy the interaction. Just as behavior and physiology guide ethological studies of wild and captive animals, so too can we observe dog behavior and dog physiology and draw modest conclusions about the functioning and reasoning mind the dog might possess. Again, however, dog thinking is just that—thinking *qua dog*—and to over-import our ideas or terminology into the story would be to tread too deeply in anthropocentric waters. Appreciating the specific lifeworld of a dog is key to approaching an understanding of the specific dog mind.

There is a further consideration that I will argue demonstrates just how different canine minds must be from our own, and hence why we must continue to avoid overly liberal anthropomorphism. What I shall discuss, however, also points to an inextricable link between the mind of a human and the mind of a dog, and thus the impossibility we face trying to study dog minds in complete isolation. Instead, many of the studies surrounding dogs and how they think are better understood as queries into how dogs and humans think-*with* one another, or what I have termed "coactive cognition." This is because from the beginning—and I mean from the beginning of the evolution from *Canis lupus* to *Canis familiaris*—dogs have responded *to*, learned *from*, and thought *with* their human companions. Indeed, it was this cooperative relationship that helped transform some wolves into the loyal companion that might be sitting with you as you read this. This transformative relationship, I will argue here and throughout the rest of the book, is bidirectional. As much as humans worked to shape the wolf-become-dog mind, so too did this relationship change us, as it continues to do today.

To understand what coactive cognition is supposed to be, it is helpful to turn back to the previous section, where the "radicals" of cognitive science were discussed. Specifically, the enactivist view, which claims that cognition is often if not always better conceived as a dynamically coupled process, fits especially well with the picture of dog minds I am sketching. Within an enactivist framework, certain cognitive processes are argued to emerge from organism-environment interactions but are not reducible to either of those constituents. As a very simple example, take reading a book. The cognitive act of reading—comprehending text as the visual system scans over it, and forming ideas of what one is reading—is clearly not possible without the environmental "prop," be it a book, e-reader, or other device. The enactivist will not necessarily go so far as to say that reading itself is occurring outside the organism,[14] but the idea is that the coupling of the reader to what is being read constitutes the process. Take away the book, and there is obviously no reading. But the reading doesn't occur *in the book* either. Rather, it emerges through the interfacing between the human and the book. There are countless other examples of how this enacted view of cognition works, but I want to turn to one in particular that highlights the type of enacted "dognition" I think best suits understanding how dogs think with us.

De Jaegher and Di Paolo (2007) describe an enactivist phenomenon they refer to as "participatory sense-making" (PSM) in order to explain how meaningful thought emerges not just from organism-environment interactions but also from organism-organism interactions. Emphasizing intentional movement, which expresses willful agency without the need of explicit linguistic representation, De Jaegher and Di Paolo argue that meaning can be generated among two or more "partners" in ways that escape traditional cognitivist explanations. Take, for example, Currie's (2007) description of a couple who have just arrived at their honeymoon suite. One of them might audibly sigh while looking out from the hotel balcony—a sigh that signals her approval and contentment, but one that simultaneously elicits a response from her partner. Her overtly perceptible reaction is a communicative act insofar as she is attempting to convey meaning—to highlight a part of the world she is experiencing with another, and to check to see if this person shares the same understanding of the experience. The orientee, her partner, does not attempt to uncover her intentions as a detached third-person observer; rather, through coordination and modulation of meaning-making activities, the intentions become readily apparent without the need for theorizing, simulating, or thinking *about* the other in an explicitly linguistic manner.

Given the above example, it is easy to envisage all sorts of other interactions that would count as genuinely cognitive, but nonrepresentational, nonlinguistic, and, most important, irreducible to one partner in a constituent pair. All sorts of findings in developmental psychology are suggestive of PSM. Joint attention in parent-infant interactions (Seemann 2011), using emotional cues as factors in decision making (Tollefson 2005), and coordinated movements leading to discoveries and shared meaning (Gilbert 1990; Reddy and Trevarthen 2004), to name just a few, all bolster the ideas put forth by enactivists like De Jaegher and Di Paolo. I have argued in a paper (2015b) that certain forms of dancing—in particular, contact improvisation—also demonstrate the role movement plays in thought, and how it can actually be counterproductive to rely on an internalist, representationalist, and/or overtly linguistic strategy. Another arena in which I think PSM can be applied is in characterizing the ways in which dogs think. This is because dogs almost always think *with us*.

Donna Haraway (2008) has recently described our relationship with dogs in a way that mirrors the enactive view of cognition generally, as she claims that dog-human coupling constitutes a mode of thought that emerges in the here-and-now interactions between us and our canine companions. Focusing on the sport of agility, she argues that this activity highlights the irreducibility of thought to either of its component parts. For the reader unfamiliar with agility, it is a competition of person-dog pairs wherein the dog is cued to run through a predetermined series of obstacles—hoop jumps, teeter-totters, tunnels, weave poles—and often, though not always, the fastest dog to complete the course wins. The shared meaning between my dog and me as we navigate an obstacle course, as Haraway describes it, is a genuine communicative act, but one that no linguistic or representational account of cognition could properly capture. Though I might speak to my dog—commands like "Weave!" or "Chute!"—the bulk of the communication is going on nonverbally. Hand cues are very important, as are head nods, and my overall posture. More important, I am fairly certain my dog does not understand much English, other than the few words he has learned to associate with various actions. Yet the fact that I can simply nod in the direction of a tunnel and he runs through it and pops out on the other side, donning what seems to be the biggest dog grin ever, eyes intently staring into my own as if to say, *I did good, right?*, makes it pretty difficult to deny that communication is taking place. However, if I were to assume he understood the course in the exact same way I did—what the word "chute" represents, how faster equals better, rewards, disqualifications—then I would be going too far. Here is an example of how I can appreciate the unique umwelt of my dog and how vastly different it is from my own, despite our being on *the same course*. He is poised differently, on four legs, is using his nose more than I could ever imagine, sees the world at about a third of my height, likely in different hues, and does not grasp all the symbolic referents in this shared space to which I am privy.

Nevertheless, despite living in different umwelten within the same physical world, my dog and I, in this moment working through an agility course, are also not entirely distinct and incomprehensible to each other. To relate all of this back to participatory sense-making, agility creates a domain of shared meaning—I would go so far as to call it *social sense-making*—in which understanding is generated by interactions. My dog and I are thinking through the obstacle course together, such that our shared meaning is irreducible to either one of us alone. Much as when I consult my GPS for directions and the belief I form about my intended destination is constituted by the coupling of me and my tool, so too is this special form of thinking that exists between my dog and me.

While the sport of agility provides an excellent case of human-dog social sense-making, we need not confine ourselves to such a specialized activity with which not everyone is familiar. Instead, coactive cognition occurs among even the most seemingly mundane of human-dog interactions. Hare's studies show that dogs use social referencing to extract meaning from us, as the pointing and gaze-following experiments mentioned earlier indicate. Thus, every time you glance out the window expectantly and your dog follows along, perhaps growling or perking up its ears, you and your dog are participating in shared meaning-making. Likewise, as Berns (2013) and others have been showing, simply looking into our eyes is a mode of communication for our dogs; it is often a way of "hugging us with their eyes," as Hare describes it. And of course, as many dog trainers will tell you, along with researchers like Horowitz (2010), the way humans carry themselves—on walks with their dogs, around the house, around others—is of utmost importance for the dogs' comprehension of what the humans are thinking. Braitman's groundbreaking 2014 book on animal psychoses and emotions is perhaps most telling in this regard; our dogs appear to be capable of discerning when we are sad, angry, happy, or anxious. Based on the experimental findings I've discussed so far, dogs are likely extracting this information from our facial gestures, from our voices, and from the olfactory information they obtain from us. Even the way we move might convey information to dogs about what we are thinking and feeling.

At this point it could be argued that I am guilty of the same unwarranted anthropomorphism I warned against in the beginning of the book. Indeed, I am helping myself to the interpretation that dogs *comprehend* what we are thinking and feeling, when in reality they could simply be observing behaviors and generalizing on the associations drawn from those observations, using their "data" to determine how to act or behave. I will address some of these worries in more detail in the next chapter, but it is worth pausing briefly here to defend against this objection, which is what Rivas and Burghardt (2002) might call *anthropomorphism by omission*. This is a sort of appeal to ignorance, wherein we anthropomorphize and fail to take into account an important potentially confounding variable—the animal's particular umwelt—and how that might afford the animal very different experiences and perceptions. Likewise, if as I just described, behaviorism can explain the dog's behaviors without appealing to anthropomorphic mentalism, should we not adopt that framework?

One response to these worries is to cast them back onto the human-

human folk psychologizing that is so commonplace when we try to decode the actions of others. When we reason about the behaviors, facial expressions, and gestures of other humans, although we might not be guilty of anthropomorphism by omission—since we arguably do understand and live within the same umwelt as other humans—we could still be mentalizing when we need not be. In other words, when I see a colleague with a furrowed brow and squinting eyes staring for some time at a projector screen, rather than assuming this person is "perplexed" or in disagreement with what is presented, a simpler explanation might be that this person is scanning the environment for more information to guide future actions. The same problem of uncritical folk psychology applies to the ways humans interpret one another. However, as I have been arguing so far, completely abandoning folk psychology in favor of a strict behaviorism is likely to prove unhelpful. Not only is it possible, as Allen and Bekoff (1997) have argued, that folk psychology can be a fundamentally important part of rigorous scientific investigation, but they also point out that folk psychological frameworks often prove to be far more eloquent and explanatorily parsimonious than the alternatives. Consider, for example, the connectionist approach, which might say that we can explain behavior as a "vector transformation" from input vectors, to those that guide things like muscle contraction, movement, vocalization, and other outward behaviors. This type of account jibes well with the strict behaviorism we have been discussing, and it provides an explanation that avoids any internal mentalism or semantic content. But, as Allen and Bekoff argue, an account that completely disregards mental content does not fit well within the ethological paradigm, because such content-denying theories fail to explain the *appropriateness* of the behavioral responses generated by these "vector transformations." Connectionism might provide an explanation of what causes the responses, but not *why* those responses are better or worse suited for this or that interaction. Cognitive ethology is tasked with more than simply the "how" of animal behavior, but with the "why" as well. Thus, as Allen and Bekoff go on to argue:

> A theory that describes the internal states of organisms in terms of content seems better equipped to provide explanations. That is, the explanation of why certain cognitive states are adaptive is more complete if those states are understood to have content relating them to the environment of the organism. Of course, it might be argued that this extra explanatory value is illusory, and that it only seems as though ascribing content gives us better explanations because, for example, it makes talking about the cognitive states

easier. However, if content-based descriptions are in fact less cumbersome than syntactic or behavioristic alternatives, this provides them with a *prima facie* edge. Thus, all other things being equal, cognitive ethologists have an interest in making use of intentional notions. (1997, 73)

What, then, of our ability to "read" dogs? Surely we risk engaging in uncritical anthropomorphism if we immediately assume that a wagging tail indicates happiness or that alert eyes and ears mean excitement. But if we ascribe internal mental content, such as happiness, to a dog as a means to explain why this behavior is only elicited in certain contexts, then this is an example of the usefulness of folk psychology Allen and Bekoff defend. If we furthermore take care to catalogue and describe tail-wagging varieties, to be sure only this specific type co-occurs with the mental state of happiness, and so forth, then we can be doubly sure not to be engaging in uncritical anthropomorphism. Moreover, assuming we are also taking into consideration the umwelt of the dog, how the dog has evolved in relation to that life-world, and, most important, *why* this behavior might have evolved as an appropriate response to the given context, then we avoid anthropomorphism by omission as well.

It is my contention in this chapter that adopting an enactivist framework will provide the best means for navigating the special ways in which dogs think alongside humans while avoiding both egregious anthropomorphism and uncritical anthropocentrism. If we treat dog cognition as co-constituted with human thought and behavior, then on the one hand it would be unhelpful *not* to retain at least some human-centered ideas pertaining to thought—after all, we are half of the story. On the other hand, we are *just half* of the story; we must also describe and explain the interactions between humans and dogs, and moreover appreciate the unique umwelt and cognitive architecture of the dog if we are to successfully understand the canine mind. Now that we are equipped with growing scientific research indicating that humans and dogs do in fact possess many of the same neural processes in response to similar stimuli and hormonal reactions within our blood, it is not an overzealous move to claim that a lot of what goes on in human cognition resembles what goes on in canine minds.

This is also true when it comes to affect, but as we discussed earlier, affect is not simply internal "feelings" or states of a singular system. Affect includes social and background emotions, which are both arguably intersubjective. Evidence suggests that dogs experience these more intersubjective and wholly embodied emotions. Laurel Braitman's 2014 account of everything from canine anxiety disorders to forgiveness by whales is an excellent summary of the most current trends in understanding moods in animals. There is even reason to believe that animals might become suicidal and carry out plans for self-harm, though these claims are highly contested. What is hardly contentious, however, is that we are not the only creatures who experience bouts of deep depression or lasting compassion. The thing that sets dogs apart from other animals, again, is that these background emotions almost always operate in conjunction with humans and are experienced synchronously with them. Dogs, for example, remember their owners even when they have been gone for years, and there are countless tales of dogs sinking into despair upon the permanent loss of an owner, dogs who traverse entire continents to find a lost companion, and dogs who will put their own lives at risk to save the life of a human.[15] Braitman's own dog Oliver suffered so much anxiety over her absence that he plunged through a plate-glass window and fell several stories to the street outside her apartment. Though this incident did not kill Oliver, he would eventually go on to chew and swallow large chunks of the kennel in which he was being kept. He once did this while Braitman was out of town, and it resulted in bloat, a typically fatal condition in dogs, which was sadly the last straw for Oliver.

Dogs need humans, and not just for food and shelter. To be sure, the beginning stages of wolf domestication were marked by food scarcity, hunting demands, and survival concerns, but as dogs continued to morph into the myriad breeds we know today, they began to love us and desire our love in return. This happened in tandem with the development of highly specialized mindreading and social referencing skills. But it wasn't just dogs who benefited from this bonding process. Humans have gained seeing-eye dogs, military service dogs, seizure detection dogs (Howbert et al. 2014), and even dogs who can sniff out cancer (Willis et al. 2004). Dogs are so attuned to us that they often know things about us we have yet to discover ourselves. I suspect that many humans have similar bonds with other humans—long-term partners, best friends, close family members. It is not far-fetched to claim that others often know us and can predict our thoughts and feelings with somewhat alarming accuracy. I think that this attunement to the other, demonstrated most clearly in instances involving shared background emotions and social affectivity, is itself a sort of backdrop against which both the humans and domestic canines developed alongside one another. This backdrop, I argue, is precisely the type of ecological niche many cognitive scientists have claimed is responsible for the unique and varied forms of intelligence observable in the animal kingdom. When it comes to humans and dogs, it is an *emotional niche* that is responsible for such highly attuned cooperation and tight synchrony (Krueger 2014). And as many of the "radicals" of cognitive science have been urging of cognition generally, these dyadic pairings between us and our dogs allow for information-rich exchanges that are not captured by appealing to an internalist or individualist account. The ecological emotions through which humans and dogs have forged a bond and shared set of affective experiences are sustained only through continued interactions between the two species.

Learning to Read One Another in Interspecies Dyads

I have argued that much of cognition—which necessarily includes affect—is better thought of in terms of intersubjective dynamics and interaction. I have also argued that this same intersubjective cognition, this "coactive cognition," is especially evident in dog-human pairings. Moreover, the bond between humans and dogs over thousands of years has itself allowed for a sort of symbiotic interdependency and a unique form of interspecies empathy. To reiterate: this symbiosis I'm arguing for is specific to those human-dog dyads who live together on a daily basis, often sharing homes, even beds. To be sure, other liminal dogs, such as street dogs in metro areas like New Delhi, might exhibit some of these same bonding skills with some humans—but the studies I've drawn on thus far exclusively pertain to dogs who interact with humans regularly and predictably. Finally, given what we know about the close connection between humans and dogs of this particular variety, I have argued that the best way to properly characterize this coactive cognition is through an enactive framework that sees interaction as preliminary to and constitutive of these shared acts of thought and emotion.

It might be insisted that the view is still anthropocentric and that, furthermore, thoughts and feelings *just are* internal. While interaction might be a cause for certain forms of cognition, the cognitive process itself is still subjective. To the first objection, I will simply refer the reader to the many discussions of anthropocentrism circulating in animal cognition literature (see esp. Andrews 2011, 2014) and state briefly that I think it is entirely unavoidable to take a human-centered view of cognition. Thinking about cognition from any but a human viewpoint is patently impossible. The real trick is to determine whether the anthropocentrism that is being employed in any given research paradigm or philosophical argument is overly liberal. To this end, I am confident that I have enough scientific evidence backing the claims I have made pertaining to dog cognition and emotion, and I have avoided equating human and dog minds as much as possible. In fact, one interpretation of my argument—indeed, the interpretation I hope most readers glean from this chapter—is that the sorts of emotions constituted in human-dog pairings are neither entirely human nor entirely canine, but are a hybrid born of the two very different species coming together and sharing experiences. One need not worry about anthropocentrism because the story is just as much about how dogs helped humans develop specific social emotions as about how we have shaped dogs' cognitive niches. Genuinely loving, caring for, and grieving over the loss of members of a species with whom you cannot even have a linguistic conversation—these are modes of affectivity that are unique to an interspecies bond and hence, I argue, not anthropocentric in how they are characterized.

As to the objection regarding the ultimate "location" of cognition and affect—namely, that they must be individual and internal phenomena—I suspect this is the heart of any objection to an enactivist or nontraditional position regarding cognition. While I stand by the argument that certain modes of thought cannot be understood subjectively, I realize this internalist-externalist debate will not be settled once and for all here. I conjecture, however, that with more empirical research and further compelling findings related to the human-canine bond, the case I am making can be made even stronger. There is reason to suspect that the intersubjective story I am telling regarding background emotions is already borne out in research, at least pertaining to human-human bonding, as ideas thought to be "fringe" or the stuff of urban legends start to find traction in empirical science. A once mythical cause of death, dying of a "broken heart," it turns out, is not so far from reality. Now termed Takotsubo cardiomyopathy (*tako-tsubo* is Japanese for an octopus trap, which is what the diseased heart resembled to those who discovered it) or stress cardiomyopathy, the condition afflicts people who suffer acute emotional distress, and it can be fatal (Virani et al. 2007; Wittstein 2007). It is most frequently seen in patients who have lost a loved one or have been through a traumatic breakup. In other words, you are more likely to suffer from Takotsubo cardiomyopathy if a sustained interactive relationship is suddenly severed. Tales in which couples are married for fifty years, then one dies and the other quickly follows, have captivated our attention in literature and movies, but the idea that your whole body could simply shut down in response to the loss of deep emotional connection to another is not just the stuff of good storytelling.

What does Takotsubo cardiomyopathy have to do with my overarching argument pertaining to human-dog relationships and enactive or ecological cognition? First, it points to the need for more research in the domain of long-term interactions between humans and dogs. Perhaps the accounts of depressed dogs and suicidal parrots discussed by King (2013) and Braitman (2014) have similar causal bases in a disruption of shared emotional bonding and loss of love. As King notes, the presence of grief in these animals implies that there was once love. Second, I think the idea that sustained and severed interactions have physiological consequences speaks to the top-down effects of these social and background emotions obtaining between humans and likely between humans and other animals. To be sure, the internalist-individualist camp can always dig their heels in and proclaim that just because interactions might result in internal physiological changes, the emotions themselves do not cease to be private affairs, hidden within the confines of the singular organism. However, if we look at the bonded pair more as a single system, at least during intense moments of interactive synchrony, the inner/outer distinction becomes harder to define, let alone depend upon. This is a theme I further explore in the final chapter of this book. But first I want to examine another domain of cognition that always involves at least a pair: social cognition. While I have mentioned studies relating to it here, I think it warrants its own chapter. Philosophers have spent a great deal of time debating what precisely constitutes social cognition—a debate in which the important skill often referred to as "mindreading" is central. Yet when it comes to how these skills are subtended in the interspecies relationship between humans and dogs, not much has been done by way of compelling research. What we have instead are studies involving mostly nonhuman primates, and a few cases in which researchers claim they have found similar mechanisms in dogs, but these are controversial cases, to say the least. Therefore in the next chapter I explore what it might mean to claim that humans and dogs engage in their own form of interspecies social cognition and mindreading. This capacity to read one another, if it exists at all, I argue, cannot be explained by standard cognitivist accounts of the mind but must instead be understood in a primarily intersubjective and enactive framework, much like the other forms of cognition and affect we have so far discussed.

CHAPTER 3

Canine Mindreading and Interspecies Social Cognition

In the previous chapter, I argued that dogs and humans engage in collaborative cognition and shared affect, and the capacity for doing so stems from the long-standing transactional history of the two intermingled species. Perhaps most compelling are the ways in which *social* emotions emerge in the dynamic interplay between humans and their dogs. Evidence of this co-attunement is surfacing in fields such as neuroscience and endocrinology (see Payne et al. 2015). Findings such as those demonstrated by Berns (2013) suggest that dogs are capable of feeling love or, at the very least, *attachment* toward their human companions.[1] As I argued in the previous chapter, not only have dogs developed the capacity for loving us, but so too have humans grown to love dogs, thanks to this coactive cognition and affective attunement. Furthermore, this cognitive and affective resonance has served both species well in terms of evolutionary fitness.

I also argued that the best way to explain how this coactive cognition emerged and has been sustained in both species is to adopt an enactivist stance on cognition, as doing so allows us to reconceive thoughts and emotions as shared, intersubjective transactions rather than individual, internal, and private events. Enactivism is also uniquely suited to the task of explaining how affect is perceived between the two species, without having to posit anything like propositional attitudes in dogs, or the ability for dogs to form theories about the mental states of others. To this end, I provided reasons to believe that in the case of understanding affective significance, one need not adopt a representational model of the mind, nor a theory that depends on conceptual content. Instead, affective significance can be *directly perceived* as it emerges in dyadic exchanges between humans and their dogs.

Perceiving affective significance is just one of many ways in which humans understand one another, and while it might be possible to read emotions directly from behaviors, this might not be true of other skills, such as recognizing goal-directedness and intention. It is far from settled whether animals besides humans possess the capacity for decoding others' minds, and if so, what that faculty is like and how it works. Indeed, how humans understand the minds of other humans is itself a matter of contention. This chapter therefore attempts to make some headway in these domains of inquiry. I start with an overview of this crucial set of skills, often referred to as "mindreading" or "theory of mind,"[2] and some of the major theoretical contenders that seek to explain this capacity, including a theory often ignored in mainstream philosophy of cognitive science: what Shaun Gallagher (2001) calls *Interaction Theory*. Interaction theory is based in the influential work of Colwyn Trevarthen (1979), who argues that what marks our first transactions in the social world is "primary intersubjectivity." Interaction theory and primary intersubjectivity go hand in hand, as I will explain. It is my contention that Interaction Theory, as it is well aligned with enactivist and ecological views of cognition generally, is best suited to explaining how we read the minds of other humans. Then I turn to the question of mindreading in nonhuman animals. Maintaining a decidedly critical lens, I examine the experimental evidence purporting to demonstrate that such skills are present in nonhuman animals and especially dogs, but I also address some of the most compelling reasons to doubt the validity and significance of these findings. Ultimately, I argue that we do have reason to believe that nonhuman animals are capable of mindreading because it is not an all-or-nothing capacity. Instead, mindreading occurs at varying levels of expertise, and an important facet of this spectrum of skills—being cognizant of *awareness relations*, or the extent to which others are aware of the same things you are—is something dogs arguably do, much like other nonhuman animals. By looking at some of the experimental findings in dog-human interactions that include play, deception, and attention, the case for a more nuanced understanding of mindreading capacity is made quite strong. Moreover, examining the ways in which humans interact and play with dogs provides further support for primary intersubjectivity or Interaction Theory as an alternative to the standard two choices—Theory Theory and Simulation Theory—in this debate. Finally, I argue that all of this taken together points back toward a more general assertion I have been defending in the book: that the enactivist view of cognition is the best candidate for explaining social cognition, both in dogs and in humans, and, more important, in understanding how human-dog interactions can constitute thought.

What Mindreading Is and Is Not

A capacity that most humans certainly possess is the ability to understand what other humans are thinking or feeling. If I see Bob pointing a finger on one outstretched hand while the other hand covers his mouth as he gasps, eyes wide open, staring in the direction of his pointed finger, most likely I will follow Bob's gaze and gesture. If I thereupon see a building on fire, I will make the quick judgment that Bob is shocked at the sight. Perhaps he is even feeling fear, wondering if anyone he knows is stuck in the building. Or he might work in that building and be worried that he will lose a great deal of his projects to the fire. All these possible interpretations of Bob's mental state as he gazes at the burning building are made effortlessly. The big question for philosophers and scientists is *how* this is done, and an even more pressing question we need to examine is *why* it is done. It might seem like a simple question with an obvious answer, namely that Bob's *behaviors* are what I am using to assess what he is thinking. And to some extent that must be what I'm doing. Indeed, as we have discussed already, there is no way to know exactly what is going on in the mind of another, be it human or nonhuman, and the problem of other minds can leave one skeptical about just how accurate we can hope to be when engaging in mindreading activities. Nevertheless, we do it every day, often without thinking much about how or why we make the judgments we do. If you were to ask me, however, why I thought Bob was in shock, I would answer that it is because he is staring at a burning building, and he looks really surprised. In other words, behavior is what we tend to rely on when ascribing mental states to others, so the seemingly obvious answer to the question of how we mindread is not wrong, per se. It is just the tip of a much larger iceberg related to mindreading research. For decades now, philosophers and cognitive scientists have been probing more deeply into the mechanisms behind our ability to take all these behavioral clues and glean from them mental significance. Is the process unconscious? Do we form a theory about what we are observing or do we simply place ourselves in the other person's mental shoes, as it were? If behaviors are the primary means by which we form these beliefs about other minds, is it possible that we use the same mechanisms to understand nonhuman animal minds? Moreover, is it possible that they might mindread as well, and if so, what evolutionary purpose would such a capacity serve? These last two questions have received a lot more attention lately, and I will discuss some of the findings and theories surrounding them a bit later. For now, let's look at the first set of questions which seek to understand the process by which humans utilize behavioral cues to understand one another.

For a long time there were two main contenders in the quest to answer how mindreading and social cognition work. Theory Theory was propounded by Leslie (1987), Baron-Cohen et al. (1985), Gopnik and Meltzoff (1996), Baron-Cohen (1997), and Gopnik et al. (1999), whereas Simulation Theory found proponents in Gallese and Goldman (1998), Heal (2003), and Goldman (2006, 2009), with hybrid versions offered by Nichols and Stich (2003) and Meltzoff (2007). While I will ultimately defend the more recent Interaction Theory, as I think this third alternative has more merit in answering some of the most difficult questions about both human and nonhuman mindreading skills, let's begin by briefly examining the long-standing debate between Theory Theory and Simulation Theory.

Theory Theory comes in several varieties, but the overarching idea is that we understand other minds by employing a theoretical stance. When I posit that Bob is worried that his work will be compromised in the fire that he is watching across the street, the Theory Theory proponent will argue I do so by first implicitly assuming Bob has mental states. I then observe Bob's behaviors in relation to the particular context, use those behaviors to theorize about the past and its impact on the present, and even make predictions about future actions in which Bob is likely to engage. Of course, all of this is done quickly and automatically, especially the part about my assuming Bob to be a person with a mind of his own. Taking a theoretical stance, in other words, does not have to be explicit or linguistic in nature. It can be as simple as a child recognizing that another child is having difficulty with a task and offering help. Well before the onset of sophisticated language skills, children and babies demonstrate understanding regarding what others are thinking. But it is precisely because, at around the age of four, children really seem to turn a mindreading corner, as it were, that proponents of Theory Theory argue this ability fully comes online, since some important developmental milestones will have been met by then. An experimental paradigm overwhelmingly relied upon in demonstrating how this capacity becomes fully functional is the False Belief Test.

Around age four, regardless of culture or language,[3] children begin to recognize that other people have beliefs and that those beliefs can be different from their own. There are many iterations of the False Belief Test, but they are all aimed at showing the same thing: until four years of age, children are not capable of recognizing that their own beliefs might be false or different from others' beliefs, which can also be false. If you present a box of crayons to a three-year-old and ask her what she thinks is in the box, she will answer, "Crayons," most assuredly. If you open the box to reveal candles instead, and then ask the three-year-old what someone else who has not seen that there are candles in the box would think, the three-year-old will say that that person would think there are candles in the crayon box. Three-year-olds will also claim that they always thought there were candles in the box, thus showing that they do not even comprehend their own false beliefs. Run this test on a four-year-old, however, and she will tell you that someone who has not seen what is actually in the box will think there are crayons in the box (Gopnik and Astington 1988). By age six or seven, children develop an even more sophisticated version of this ability—they can think about another person's thoughts about a third person's thoughts. That is, six- and seven-year-old children can explain why Julie wrongly assumed that Mary was sad because Mary was crying, when in reality Mary was crying because she had just learned that her dad was coming home early from a military tour, and Julie did not have this information.

Because the developmental trajectory of children's capacity for detecting false beliefs in others and in themselves is so uniform and predictable, many Theory Theory supporters argue that it points toward an innately specified domain-specific learning mechanism in the brain, one whose sole purpose is to generate theories about the mental states of others (Scholl and Leslie 1999; Carruthers 2013). Others argue that it is through the continual testing of hypotheses within a social environment that a child can eventually take a theoretical stance about the minds of others (Gopnik and Wellman 1992; Gopnik and Meltzoff 1996). To wade into that particular debate would far exceed the scope of this book, but it is worth noting that Theory Theory proponents often rely on a modular view of the mind. Whether they are correct or not, most agree that the primary means by which we engage with and understand others is implicit theorizing, and that this capacity develops in tandem with other important skills such as a much more advanced linguistic repertoire. While language comprehension is evident even in young infants, it is not until around age four that children are fully able to utilize their language in all its glory—talking about past events, discussing hypotheticals or playing "pretend" with language, and understanding jokes (Saffran et al. 2001; Lidz et al. 2003; Bergelson and Swingley 2012). Since these skills coincide with the ability to detect false beliefs in others, supporters of

Theory Theory generally argue that understanding other minds in general is the result of a slow but deliberate development of the capacity to take a theoretical stance toward those other minds.

The other major contender in the mindreading debate, Simulation Theory, argues that instead of forming theories about the minds of others, we generate internal models of our own mental states and use those to gauge what other people are thinking. For example, if I see someone crying while clutching a paper program for a funeral service, I do not need to take a theoretical stance, hypothesizing what it must be that is driving that person's behavior. I rely on my own internal model—I emulate the other person or put myself in that person's position—again, not by theorizing, but by running a simulation I have immediately at my disposal because I too have a mind and have experienced loss. Once the simulation has been run, I can understand what the other person is feeling because it is probably what I would feel in a similar situation. All of this can be done quickly and unconsciously, such that I am not even aware that I am running these simulations, though I will be aware of the sense of empathy I might suddenly feel upon grasping the meaning of the other person's behaviors. As Robert Gordon (1986) argues, these simulations take place internally and are representational in nature, even if they do not rise to the conscious level.

There is some compelling evidence to support Simulation Theory, such as the mirror neuron system and its role in social cognition (see Rizzolatti et al. 1996; Rizzolatti and Luppino 2001; Arbib 2002; Rizzolatti and Craighero 2004; Gallese et al. 2004). Though there is still quite a bit of debate surrounding what all the mirror neuron system is responsible for, there have been numerous studies indicating that this system of neurons is activated in tasks involving mindreading, social cognition, and language (see Rajmohan and Mohandas 2007). Rizzolatti and his team (1996) first discovered the system when studying macaque monkeys. They noticed that a particular area of the premotor cortex—the F5 region—was activated whenever the monkeys were performing certain actions and also when watching other monkeys perform those same actions. Further studies revealed that humans have a similar mirror neuron system, though it seems to be even more sensitive to the actions of others, such that when watching a goal-directed action or even just an intransitive action like jogging, the mirror neuron system is activated. Given that the system in humans is so highly attuned to human action, it stands to reason that it is at least partly responsible for some of our mindreading skills. Moreover, Simulation Theory can explain why infants seem primed to imitate facial expressions, as Meltzoff and Moore (1977) famously showed, well before any theory formation could take place. However, as we shall see, relying solely on the mirror neuron system to argue for Simulation Theory runs the risk of being overly anthropocentric and ruling out genuine cases of social cognition in nonhumans who are not endowed with the same neurobiological mechanisms.

More important, the idea that all the capacities involved in a complex skill such as mindreading could be explained solely in terms of theory formation or simulation has been questioned for quite some time now. Whether by proposing an alternative—such as Interaction Theory or a direct perception model as advanced by Chemero (2006), Gallagher (2008a), and Spaulding (2015)—or by suggesting that elements of theory formation and simulation are at play in social cognition tasks, philosophers and cognitive scientists have challenged the idea that we must choose between Theory Theory and Simulation Theory. Gallagher (2001) argues that most discussions surrounding mindreading rely on what he calls the "mentalistic supposition," which is the assumption that to know another person's mind is to know their beliefs, desires, and intentional states, all of which amounts to possessing a conceptual body of knowledge about the other person. As a corollary to this, any implicit understanding we might have of other minds is informed by this conceptual knowledge. This is true even of our own tacit understanding of our own minds, as Gallagher argues:

> To discover a belief as an intentional state even in myself requires that I take up a second-order reflective stance and recognize that my cognitive action can be classified as a belief. Indeed, to explicitly recognize that I myself "have a mind" is already something of a theoretical postulate. This is not to deny that I might have something like a direct access to my own experience, or that this experience can be characterized as self-conscious. I can easily say, for example, "I feel very good about planning my trip." But to say that this experience of feeling good is in fact a *feeling*, and that this feeling depends on a *belief* that I will actually take the trip, requires something like a reflective detachment from my phenomenal experience, and the positing of a feeling (or belief) as a feeling (or as a belief). It would involve a further postulation that such feelings and beliefs are in some fashion part of what it means to have a mind. (2001, 91)

Indeed, many of our thoughts about others' thoughts or about our own thoughts are metacognitively theorized in this way, but Gallagher argues that a great deal of our transactions with others are subtended by what he refers to as primary intersubjectivity. Whereas Theory Theory or Simulation Theory would both posit a representational and/or conceptual framework for ascertaining what another is thinking, primary intersubjectivity need not rely on any abstract theorizing, nor even implicit simulation. Instead it involves the embodied know-how we bring to bear on social situations, and, as Gallagher argues, these embodied practices are far more pervasive and underlie nearly all our mindreading capacities. Theory Theory and Simulation Theory, on the other hand, serve to explain a very narrow and specialized set of cognitive processes we engage in; for example, when we are talking to our colleagues about how best to think about metacognition. Hence, as Gallagher suggests, the more general theory we ought to adopt is Interaction Theory, and skills such as theory formation or simulation would fall under this umbrella.

Primary intersubjectivity focuses on the primacy of second-person interactions. Rather than assume other persons' minds are hidden away and must be hypothesized about or decoded, Interaction Theory says other minds are *given* in their embodied actions. Put differently, what others are thinking is expressed directly by their behaviors, whether those behaviors are intentional or not. In arguing for primary intersubjectivity, Gallagher (2008b) cites evidence from developmental psychology and neuroscience, such as the fact that prior to any ability to theorize or mentalize others, infants distinguish faces from non-faces, even when those faces are not specifically human (Johnson and Morton 1991; Legerstee et al. 1998; Farroni et al. 2005; Kato and Mugitani 2015). As mentioned above, neonates as young as ten minutes old imitate facial gestures (Meltzoff and Moore 1977), which suggests an innate mechanism responsible for imitation, one that would come online well before any theory formation or simulating. Indeed, as Gallagher has extensively and compellingly argued in *How the Body Shapes the Mind* (2005), we have an innate body schema that is responsible for all sorts of embodied action at the prenoetic level. We can "do things without thinking about them," such as how I might walk along a wooded path, actively avoiding tripping over tree roots or hitting my head on overhanging branches, all while lost in thought about something completely unrelated to these obstacles. There is an abundance of evidence from a diverse set of disciplines suggesting that a lot of the blueprints for bodily action are automatic and encoded by this body schema. And this pre-reflective capacity to act affords us social interactions as well. When young infants are able to follow gaze or pointing, thereby tacitly recognizing the bodily intentionality of the other, this is happening at the body schematic level, as it integrates received sensory information with the infant's motor systems (Gallagher and

Meltzoff 1996). And findings from neuroscience regarding the mirror neuron system support this idea—when perceiving action, the mirror neurons activate such that the observer is not really an observer at all, but is actively engaged with the other by mirroring the other's actions. As Gopnik and Meltzoff argue, "we innately map the visually perceived motions of others onto our own kinesthetic sensations" (1996, 129).

Developmental psychology and neuroscience support the idea that infants come into the world ready to perceive other minds, not by theorizing or modeling, but by interacting. They are equipped with second-person interactive mechanisms, such as face detection and mirroring, and, as Gallagher argues, "they need no internal plan to consult since they have a visual model right in front of them, namely the face of the other, as well as a proprioceptive model, namely the gesture that is taking shape on their own face" (2001, 87).

As noted, some theorists use this developmental evidence to argue for a modular view of the mind. The neonate, in this view, is born with specialized, often encapsulated, and domain-specific mechanisms. Simon Baron-Cohen (2006, 1997) argues that the mindreading mechanism is itself one of these modules, and also posits that we have an intentionality detector, an eye direction detector, and a shared attention mechanism. This would indeed help explain why infants are so quick to pick up on goal-directed action, are able to differentiate between someone gazing attentively at an object and simply turning their eyes turned toward it, and can jointly attend to that object with another person. I won't tread into the debate between the modular and non-modular views of the mind, which is not relevant to our purposes here. Whether these capacities are in fact innate and entirely encapsulated and domain-specific does not change the fact that they seem to come online extremely quickly, if not immediately at birth, and they allow for all sorts of mindreading to take place, well before the ability to theorize or mentalize about others begins.

The immediacy with which infants can recognize faces, follow gaze, and imitate others also translates to affect. As discussed in chapter 2, affective attunement and co-regulation take place in dyadic pairings between infants and adults (and people and dogs) without the need for any higher-order consciousness, computation, or otherwise "reflective" mental states. It turns out that the body schema is at work to integrate not only sensory and motor information but affective data as well. Around five to seven months, infants are able to detect correlations between visual and auditory information and the emotions that those sights and sounds convey.[4] Consequently, Gallagher argues, "before we are in a position to theorize, simulate, explain or predict mental states in others, we are already in a position to interact with and to understand others in terms of their gestures, intentions and emotions, and in terms of what they see, what they do or pretend to do with objects, and how they act toward ourselves and others" (2001, 90–91).

In short, based on the evidence from infant studies, there might indeed by a way to think about mindreading that does not depend solely on some specific version of Theory Theory or Simulation Theory. As noted earlier, Interaction Theory or primary intersubjectivity does not rule out that there are times when we might theorize about others' intentions or simulate a model in our own minds to represent what the other is thinking. But before any of those capacities come on line, our primary mode of understanding others is through *interacting* with them. Gallagher is not alone in arguing for primary intersubjectivity, though other proponents of similar approaches might term the phenomenon differently. And like many other theorists, Gallagher also argues that the idea of primary intersubjectivity is not just about the developmental story. "Primary" here means "fundamental for," not just "prior to." Our primary mode of understanding other minds, be it at infancy or into adulthood, is second-person, intersubjective interaction. Thus the argument for Interaction Theory is not only a developmental claim but a pragmatic one as well. To explain most face-to-face social interactions, Interaction Theory works quite well. Consider affect recognition, again. Derek Moore and colleagues (1997) found that subjects could identify the particular emotion expressed by actors in a dark room who were wearing pointlights at various body joints while acting out bodily movements that correspond to those emotions. This finding is reminiscent of a now famous experiment by Fritz Heider and Marianne Simmel (1944), in which subjects watch shapes moving about a screen and are asked to report what they saw. Overwhelmingly, subjects attributed mental states to the shapes, such as "the large triangle was angry at the smaller one" or "the circle was afraid of the large triangle." The experimenters in this case set out to test the hypothesis that humans will attribute mental states liberally if the movements and actions of whatever is being observed correspond well enough with those typically labeled as such. The Heider and Simmel experiment has served as a sort of warning for philosophers of animal cognition as well as artificial intelligence research: be careful how much mentality you attribute based solely on behavior, lest you engage in unreflective anthropomorphism. And rightly so—humans play exceedingly fast and loose with their attributions of mentality, as this experiment showed. But the fact that we are so quick to assume others have minds can also serve to provide further proof that something like Interaction Theory is a better way to think about the mechanisms underlying our mindreading capacities. As experiments like those conducted by Moore et al. show, movements, even by point-lit silhouettes of humanlike shapes, provide ample information for us to detect the emotions expressed. Much as I argued in chapter 2, therefore, emotions are often just given in the facial expressions or gestures of the other. Likewise, what someone is thinking, planning, trying to accomplish, or mad about is made directly available in their embodied actions and movements.

I have provided a concise account of the debate between Theory Theory and Simulation Theory in explaining the mindreading capacities of humans and why I think a third alternative, much like Gallagher's primary intersubjectivity or Interaction Theory (see also De Jaegher et al. 2010), is better suited to explaining our primary way of understanding others. There are certainly worries to entertain with regard to this view, and I intend to address some of those concerns, but rather than doing so at present, I want to turn back to the central focus of the book, our canine companions. I think exploring the ways dogs may or may not be capable of the same mindreading skills humans possess can shed even more light on the mindreading debate. As it turns out, Interaction Theory is also well poised to explain the ways in which humans and dogs interact with and understand one another. In fact, as I will argue, it may well be the *only* way to explain "canine theory of mind," if such a thing exists. Thus it is my hope that by examining the ways mindreading takes place between humans and dogs, many of the objections that might be levied against Interaction Theory are rendered moot, given the even stronger pragmatic force of the arguments in its favor.

The Social Cognition of Canines

In this section I reexamine some of the findings discussed in previous chapters that purportedly indicate dogs are able to decode at least some of what is going on in the minds of humans. Additionally, I discuss some empirical evidence in favor of the idea that dogs might also understand the minds of other dogs. In my view, there is sufficient reason to believe that the study of canine cognition ought to include social cognition among its topics, as at least some form of mindreading is demonstrated in a variety of experimental findings. However, before reaching this conclusion, I will first pause for some methodological considerations; most important: *Is it even possible to measure mindreading in species other than humans?* My answer ultimately is positive, with two caveats. First, *what* we are measuring and *how* we are doing so in canine social cognition might turn out to be very different from what we find and how we find it in humans. Second, I propose that mindreading specifically and social cognition generally ought not to be thought of as an all-or-nothing set of traits but rather as a gradient or spectrum of capacities, such that canines might possess some of those traits and not others, while other species, such as apes, might possess a different set, and so forth. Once I have defended this gradient view of mindreading, I return to the debate between Theory Theory and Simulation Theory to argue that neither framework is suited to explaining canine social cognition. These theories do not suffice to explain the mindreading that occurs either *intraspecially* between two or more dogs or *interspecially* between humans and dogs. Instead, when we are measuring the mindreading capacities of dogs, our best bet is to rely on something akin to Interaction Theory. I conclude that mindreading is not one singular skill, but instead ranges from full-blown representations of the knowledge and ignorance of conspecifics to simply understanding "awareness relations" (Martin and Santos 2016), and that the complex feature of cognition more generally thought of as "social cognition" cannot be neatly explicated by any one simplified theory.

Is Mindreading in Nonhuman Animals Empirically Tractable?

Before delving into the experimental work purporting to show that dogs and other nonhuman animals have mindreading capacities, it is worth asking a more fundamental question: Is it even possible to measure mindreading in nonhumans? Given the considerations discussed thus far, answering this question will depend largely on (1) what exactly we mean when we say an animal is "reading the mind of another animal" and (2) whether this definition is consistent among different species. For example, it might be an obvious indicator that Bob has a theory of mind if he is able to pass the False Belief Test, but this is surely not the only measure of mindreading, and if we set a dog the task of detecting false beliefs in the same manner humans are typically subjected to the test, the dog will certainly fail. Yet this does not rule out the possibility that there are other ways to demonstrate some sort of mindreading abilities in dogs. Let us therefore discuss what it would take to design an experiment that actually tracks what we are hoping to measure— because, as it turns out, many of the findings that purportedly demonstrate nonhuman animal mindreading are deeply flawed methodologically.

Ever since David Premack and Guy Woodruff (1978) embarked on trying to determine whether chimpanzees recognize intention and goal-directedness in their human caretakers, countless studies have attempted to do the same, with all manner of animal, from corvid to canine. One well-known experiment by Hare et al. (2001) placed chimpanzees in a "food competition" setup, in which a dominant chimp and a subordinate chimp were facing each other and had food placed between them. Depending on the configuration, the dominant chimp could either see or not see where food was hidden, and when the dominant chimp clearly did not see where the food was placed, the subordinate chimp usually attempted to take it. The subordinate chimp avoided trying to obtain food that the dominant chimp saw being hidden. From these findings, Hare et al. concluded that chimpanzees understand the mental states of other chimpanzees because they recognize when something is "seen by" or "not seen by" the other and can form the belief that the other does not know food exists if it is not seen.

Experiments like the one just described are criticized from several angles. José Bermúdez (2003) would argue that the experiment proves nothing other than that behavioral associations based on learning are taking place. For Bermúdez, the question of whether nonhuman animals have mind-reading capability—or theory of mind (ToM), his preferred term—can be answered a priori because, according to him, in order to possess ToM, one must be able to have *second-order thoughts*. In other words, one must be able to think about thoughts. And to be able to think about thoughts, one must be able to represent those thoughts symbolically in a natural language. Another way to think of it is that for thoughts to have what Bermúdez calls "intentional ascent"—that is, to play an inferential role in guiding beliefs about the thoughts of another—they must have "semantic ascent": they must, in other words, be linguistic. Since nonhuman animals do not have language, ipso facto they don't have ToM.

Bermúdez's argument certainly rests on some contentious premises, and philosophers like Robert Lurz (2009, 2011) are quick to point those out. In particular, the assumption that higher-order thoughts require that the vehicles of those thoughts be (1) at the personal level and (2) linguistic, according to Lurz, raises a question about consciousness, namely: If the second-order thoughts are conscious (and presumably they are, if an animal is "aware" of another creature's mental states), then the vehicles of those thoughts—the personal-level representations of the other's thoughts—must also be conscious. However, this is a huge assumption to pack into an argument, given that it is quite possible that the representations could be carried out at the sub-personal, nonconscious level and then give rise to the second-order, conscious thoughts. In other words, as Kristin Andrews notes, Bermúdez is denying a distinction between first-order thought and metacognitive thought, a "distinction that many take to be essential" (2014, 145). David Rosenthal (1993, 2004) has argued extensively about this issue, and there is nothing in Bermúdez's argument that shows why we ought to believe that the vehicles of metacognitive thought must themselves also be conscious.

The debate between those who side with Bermúdez and those who side with Lurz is philosophically engaging, and I think a great deal rests on it, but I don't want to spend too much time with it because I think an even more pressing question is at hand: *Must all thoughts about thoughts take place in language?* Of course, another question looms large—and its answer is one that Bermúdez helps himself to with very little defense—which is the question of whether nonhuman animals are incapable of possessing language in the first place. That too is beyond the scope of this book, but it is worth noting that Bermúdez is echoed by many in claiming that humans are the only species with language. Influential thinkers like Chomsky (2006) have convincingly distinguished between communication and "natural language," and so perhaps Bermúdez is right to simply assert that premise. My contention is that it doesn't matter either way, because the answer to the first question I posed—whether all thoughts about thoughts must be linguistic—is ultimately no. I will return to this claim shortly, but for now let's assume for the sake of argument that I'm correct. Indeed, this would be the only way to even begin trying to test for mindreading capacities in nonhuman animals, if Bermúdez is correct in assuming that only humans have language. We would have to test for mindreading in a way that does not rely on linguistic data. So is *that* something we can hope to do?

Derek Penn and Daniel Povinelli address this question in their paper "On the Lack of Evidence That Non-Human Animals Possess Anything Remotely Resembling a 'Theory of Mind'" (2007b; see also 2007a). The title alone gives away their thoughts as to whether nonhuman animals can mindread. Despite their conclusion that, indeed, no nonhuman animals exhibit anything that would suffice to prove they have mindreading capacities, Penn and Povinelli do provide a useful framework for designing an experiment that in principle would track the capacity. There is a lot of formalization to their requirements that need not be repeated here, because the general constraints are fairly easy to capture. In short, we need to make sure that the experiment is designed in such a way that the very best explanation for the observed behavior or action or whatever we are measuring is that it is in fact a mindreading mechanism at work and not some other, simpler function. According to Penn and Povinelli, "One must, in other words, create experimental protocols that provide compelling evidence for the cognitive (i.e. causal) necessity of an f_{ToM} *in addition to* and *distinct from* the cognitive work that could have been performed without such a function" (2007b, 734).

This echoes an earlier paper Povinelli wrote with Jennifer Vonk, in which they directly refute the validity of Hare et al.'s 2001 chimpanzee food competition experiment on the grounds that the chimps' behavior can be explained in terms of the "behavioral invariants (looking, gazing, threatening, peering out the corner of the eye, accidentally spilling juice versus intentionally pouring it out)" (2003, 159) to which the chimps have access and from which they learn over time. Povinelli and Vonk argue that an experiment needs to not rely on the animal's past experiences with other animals' behaviors, but should instead involve the animal having to "improvise" and draw inferences based on new information.

With these requirements in mind, again, the question remains: Is it possible to devise such an experiment? Penn and Povinelli, along with Povinelli and Vonk, do think it's possible, and they cite a proposal by Cecelia Heyes (1998) to expose chimpanzees to a completely novel situation in which they indeed could not rely on past interactions or stored information. In the Heyes experiment, chimps would be given the opportunity to wear two different types of buckets on their heads—one type with a see-through visor and the other with an opaque visor. The buckets would be randomly color-coded and sorted, so no other physical attributes of the bucket could be mistaken for the cue that would tell the chimps that one bucket is see-through and another is not. Then when humans, who have food with them, place buckets on their own heads, experimenters would wait to see which human the chimps beg from. According to Heyes, as well as Povinelli and Vonk, if a chimp has coded the first-person experience of wearing a "seeing" versus a "non-seeing" bucket, and can then attribute an analogue of this experience to the human participants now wearing the bucket, this would count as the chimp using information about the mental states of others to determine from whom it is best to seek food.

It is worth noting that even the experiment proposed by Heyes could be subjected to the same criticism that is levied against so many supposedly compelling cases of mindreading in nonhuman subjects. As Andrews (2005) argues, we could explain chimps who beg more from the "seeing" human not as generalizing from their own mental experience but as generalizing from their own physical experience. After wearing a bucket with an opaque visor and being unable to walk around without running into things, the chimp could then apply this same procedural knowledge to the person wearing the opaque visor; the chimp sees that the person is constantly bumping into things, unable to reach out and grab the objects sought, and so forth. Thus the chimp might make the assumption that it is best to beg only from the person who is capable of accomplishing basic behavioral tasks, and this has nothing to do with mindreading.

I think Andrews's concern is important, although I think it can ultimately be dispelled by considering that the very same problem could apply to human subjects put in a similar experimental paradigm in the position of the chimps—that is, having to determine, in a novel situation, from which human it is most pragmatic to ask a favor. In the same way that the chimps' performance on the task could be characterized in terms of physical information about what the humans can and cannot do, so too could human performance, when engaging in a theory-of-mind test, often be explained in purely behavioral terms that posit nothing about inner mental states. I will not say more about this point just yet, as I plan to return to it when more seriously engaging in what the findings suggest regarding canine theory of mind, but suffice it to note that the "logical problem" inherent in any nonhuman animal theory-of-mind test is arguably present in many human theory-of-mind tests—namely that it is nearly impossible to determine if what we are measuring is a genuine case of mindreading and not behavior reading. Or perhaps it truly is behaviors all the way down, and in that case the objections just discussed are moot.

Another concern I have about experiments such as the one Heyes proposes is that it's not altogether clear, just because the chimps beg from the "seeing" person, that the chimps view that person as more willing to offer a treat. Moreover, even if the chimps figure that the person's mental state is something like "more willing to offer a treat," this might not be causally connected to whether the person can see. Consider the experiments by Brian Hare among others (Miklósi 2014; Miklósi and Kubinyi 2016) where dogs are asked to abstain from taking a treat. Depending on whether the human involved is looking at the dog or not, the dog will be either inclined or disinclined to obey. Humans with their backs turned to dogs tend to be disobeyed—the dog will take the treat despite being told not to—while humans that are looking at the dogs tend to be obeyed. The same effect holds, to a slightly lesser extent, when the human is still facing the dog but with eyes covered. What these experiments suggest about canine mindreading is debatable, but what they highlight related to the bucket-on-head chimp studies is that we must not overlook the possibility that perceiving someone as "not seeing" is cause for an animal to choose the person as more likely to provide food, perhaps because the person is seen as more easily tricked and/or gullible. We will return to these considerations soon.

The more salient point here is that, according to skeptics such as Penn and Povinelli, in experiments like the one Heyes suggests, no nonhuman animal ever successfully demonstrates possession of a theory of mind. Instead, the mechanism that has been purportedly observed is just a recharacterization of actions that are based on access to the behavioral invariants of the conspecifics. Put simply, what we always have are cases of *reading behavior* and not *reading minds*. To claim otherwise is to commit a nominal fallacy: as we know, simply naming something does not explain it.

I believe there is good reason to reject this conclusion, not so much because the argument is inherently flawed, but because several of its central premises rest on assumptions that are nowhere near settled in the philosophy of mind and cognitive science. The argument, according to Penn and Povinelli, and others like Bermúdez, is roughly this:

1. In order to engage in mindreading, one must be capable of forming thoughts about another being's mental states.
2. Forming thoughts about another being's mental states requires one to possess thoughts about one's own thoughts (what some, like Bermúdez, refer to as second-order thoughts).
3. The mental states of the other plus one's second-order thoughts about how one's own thoughts connect to action must be the necessary and sufficient cause of any assessment one makes about the other's actions. This means that there should be no other explanation capable of doing the same work.
4. There is, however, a competing explanation for the "mindreading-like" behaviors observed—namely, behavior-reading.
5. Therefore, no evidence should convince us that nonhuman animals can engage in mindreading.

So the argument is valid as stated, but premises 1–3 are packed full of assumptions that need to be teased out a bit more. Rather than doing that via traditional philosophical methodology, I want to examine some of the supposedly unconvincing evidence from research on canine mindreading. Doing so, I argue, provides excellent concrete examples of reasons to reject much of what premises 1–3 are assuming.

Evidence for Mindreading in Nonhumans?

It has been two decades since the discovery that dogs extract information from human pointing (Hare et al. 1998; Miklósi et al. 1998; Hare and Tomasello 1999). Since then, confirmation that dogs are indeed attentive to human pointing has come from numerous variations of the original studies conducted by Soproni et al. (2001), Hare et al. (2002, 2010), Miklósi and Soproni (2006), Wynne et al. (2008), Dorey et al. (2009, 2010), Gácsi et al. (2009), Wobber et al. (2009), Helton and Helton (2010), Udell et al. (2010a, 2010b, 2014), Kraus et al. (2014), Udell (2015), Zaine et al. (2015), and MacLean et al. (2017). While the experiments differ in some important ways, the general idea is the same as the one we examined in chapter 2: if dogs are unsure of where a treat is hidden, they are overwhelmingly likely to rely on the gesture—whether made with a finger, whole arm, foot, or even gaze—of their human counterpart to determine the location of the treat.

These experiments alone do not prove that dogs have a theory of mind. In fact, as Robert Mitchell and colleagues point out, these studies tend to overlook the fact that many dogs *do not* use pointing to extract information, and thus we cannot even make the claim that this ability exists species-wide. Hare claims we ought to just ignore the dogs who do not follow pointing gestures, as we ought to be interested in the ones that do, why they do, what it means, and so on. But Mitchell argues that the dogs who do not follow pointing gestures are equally important to the research. For one thing, it is possible that the dogs do in fact understand the gesture but are resisting or refusing it. Resistance to following cues might tell us a lot more about a subject's ability to understand the intentions of others than obedience to those cues. We will return shortly to this idea of resistance and how it shows up in interactions between dogs and humans in play. What is important now is to recognize that despite the significance of the findings regarding dogs' abilities to follow human pointing (indeed, primates have not shown nearly as much skill in this regard, so it continues to impress researchers to find that dogs, with little to no training, just seem to naturally do this), we must also keep in mind that (1) this is not a species-level capacity, and (2) the ability does not prove that dogs are thinking about the thoughts of humans.

Nonetheless, Hare is correct to claim that abilities such as following gaze and pointing—what are collectively referred to as "social referencing" skills—are foundational for language and culture, and arguably are necessary if not sufficient skills one must acquire in order to begin decoding the minds of others. Ultimately, my argument is going to rest on these foundational aspects of social cognition, as I put forth the idea that mindreading is not a univocal capacity, but rather exists on a spectrum. Still, we must exercise caution in jumping on the bandwagon to claim that dogs possess mindreading skills simply because some dogs in some labs follow some pointing gestures to find food.

Healthy skepticism is useful for examining purported evidence of mindreading skills in other species as well. Besides canids, researchers have claimed to demonstrate mindreading or a theory-of-mind mechanism in apes (Hare and Tomasello 1999; Flombaum and Santos 2005; Kano and Call 2017), corvids (Bugnyar and Heinrich 2005), elephants (McComb et al. 2000; Plotnik et al. 2006), and dolphins (Tomonaga et al. 2015). In corvids such as ravens, crows, and scrub jays, for example, food-caching behavior has been studied extensively, with researchers purporting to show that these birds understand when other birds are able to see them and will hide food appropriately (Dally et al. 2004; Clayton et al. 2007). Crows have been shown to memorize and recognize human faces over long periods of time (Marzluff et al. 2010, 2012), suggesting they are extracting social information from humans in some of the same ways dogs do. Even fish have made it onto the scene of ethological arguments for the existence of nonhuman animal mindreading (Balcombe 2016).

Recall, however, that Penn and Povinelli do not think such experiments demonstrate mindreading in nonhuman animals. Specifically in the case of the corvid studies, they say: "In all of the experiments with corvids cited above, it suffices for the birds to associate specific competitors with specific cache sites and to reason in terms of the information they have observed from their own cognitive perspective. . . . The additional claim that the birds adopt these strategies because they understand that 'The competitor knows where the food is located' does no additional explanatory or cognitive work" (2007b, 736). Even when considering the test that Penn and Povinelli do believe could suffice for demonstrating mindreading in nonhuman animals—the Heyes bucket test with opaque/clear visor—no animals effectively pass this test, so there remains no conclusive evidence of nonhuman animals possessing theory of mind. Vonk and Povinelli (2011) ran the test on chimps who were extremely enculturated to humans, and they failed, while eighteen-month-old human infants pass (Meltzoff 2007). Thus Penn and Povinelli think that, at best, what we can find evidence for are sophisticated behavior-reading capacities in apes, corvids, and many other species, but not genuine mindreading.

When it comes to the featured species in this book, *Canis familiaris*, the study first suggested by Heyes in 1998 and approved by Penn and Povinelli as a valid test for mindreading has not been carried out per se, though a similar and quite intriguing experiment by Monique Udell and colleagues (2008) has been carried out with both dogs (*Canis familiaris*) and wolves (*Canis lupus*). In this experiment, which also mimicked the primate begging paradigm first put forth by Povinelli and Eddy in 1996, four groups of canids were tested to see what preference, if any, they would give to begging from humans in different situations. Those situations were (1) companion dogs tested inside, (2) companion dogs tested outside, (3) tamed wolves tested outside, and (4) nonadopted shelter dogs tested in the shelter in which they were accustomed to living. Humans were either "attentive" (able to see) and looking toward the canids, or they were "inattentive" (blind), which meant they had their backs turned to the canids, or buckets over their heads, or either a book or a camera in front of their faces. An initial reading of the results from this experiment might seem bleak for anyone hoping to have found evidence of canid mindreading, especially when considering Penn and Povinelli's strict criteria. All groups of canids tested well above chance in the inattentive-back-turned situation. In the inattentive-book situation, pet dogs inside and pet dogs outside were the only ones to perform above chance. There was no statistically significant difference in begging preference among any of the canid groups between the attentive and inattentive-bucket or inattentive-camera scenarios. Only 7 of the 32 subjects preferred the attentive human over the bucket-wearing one, for example.

Udell et al. conclude from their findings that an additional mindreading mechanism is not necessary to explain the performance of the canids, although the experiment does not, *in se*, rule out the existence of such a mechanism. They also argue that the domestication hypothesis championed by so many canine researchers—namely, that domestication is either necessary or sufficient for the specialized capacities observed in dogs—is flawed, because in this case dogs with questionable human involvement (shelter dogs) and dogs with minimal human involvement (wolves) performed the same as pet dogs in the inattentive-bucket and inattentive-camera situations. Furthermore, all groups improved tremendously in the ability to choose the attentive human with very brief training, suggesting that the ability is not something dogs have evolved over long periods of time, but rather is a skill that can be learned quickly with interactive guidance. Hence, positing a special "mindreading mechanism" innate to dog minds is highly untenable.

What conclusions are we to draw regarding these discussions thus far? I don't think it's as simple as claiming dogs do not possess the cognitive architecture necessary for theory of mind and therefore are incapable of mindreading. Rather, I think the experiments on canine social cognition, much like some of the experiments on other species, point to a wide variety of skills that can be acquired by some members of the species, and the extent to which these skills are acquired is highly dependent on interaction, either among other members of that same species or, in the case of dogs, interaction with humans. Consider the fact that companion dogs, both indoors and outdoors, discriminated between the attentive and inattentive-book scenarios better than the shelter dogs and wolves. In a response to Udell et al., Alexandra Horowitz (2009) suggests that findings such as these indicate that perhaps we ought to take more seriously those who have sought to place mindreading on a continuum, rather than seeing it as an all-or-nothing capacity. Even though the findings of Udell et al. were not statistically significant enough to meet Penn and Povinelli's strict requirements for mindreading, it is not insignificant that those canids who had spent the most time around humans and had become accustomed and attuned to their habits were definitely better at picking up on subtle cues like attentiveness and distraction.

Martin and Santos argue that even nonhuman primates do not qualify as possessing the representations necessary for mindreading, because in order to pass the False Belief Test, they must recognize that another primate is in a state of ignorance. This "requires an organism to form a relation between an agent and a state of the world that is, in an important sense, decoupled from the organism's own reality" (2016, 379). In several studies cited by Martin and Santos, nonhuman primates fail to meet these requirements, as indicated by their inability to act and make predictions based on what another agent is unaware of, as well as their inability to purposefully construct situations that would make another agent ignorant (for instance, of a high-quality food item's location). However, Martin and Santos do not go on to conclude that these primates possess zero mindreading skills. Instead they point out that successful performance on some theory-of-mind tasks—making predictions based on another agent's ignorance of a food item's location, for instance—suggests the primates can form "awareness relations" whereby they comprehend what another primate can or cannot see in a given situation and use that information to determine action and make predictions. Applying this same framework to the canine experiments thus far discussed, it works well to explain the findings showing that some dogs will "disobey" and steal food when a human is not looking. They have formed the appropriate awareness relation to the human such that it is evident in that interactive moment that the human does not have awareness of the treat's location. Or, in the experiment that Udell et al. argue shows that dogs probably have no dedicated theory-of-mind mechanism, while this conclusion might be true, it does not further imply that the dogs are incapable of forming awareness relations—of understanding, for example, that the human whose view is blocked by a book is unaware.

All of this suggests that mindreading involves a plethora of skills and is not an all-or-nothing capacity. Thus, much like Horowitz, I will argue that mindreading is best thought of as a continuum, ranging from nonexistent to complex and representational. And as Martin and Santos argue, somewhere in the middle of that spectrum is the ability to form awareness relations, which I think maps nicely onto what Horowitz calls "rudimentary" mindreading.

Mindreading on a Spectrum

The staunch skepticism regarding nonhuman animal mindreaders that comes out of Penn and Povinelli and such detractors arguably stems from what is meant by "theory of mind" or mindreading in the first place. Indeed, as I will argue, what is typically meant by the term "thinking" is itself laden with contentious assumptions and likely to blame for many of the disputes over what animals are capable of and what they are not. For example, Penn and Povenelli state at the outset of their attack on nonhuman mindreading: "To begin, let us agree without too much argument that cognitive agents—biological or otherwise—can learn from their past experience, in part because they have dynamic internal states that are decoupled from any immediate physical connection to the external world" (2007b, 732). This "uncontroversial" claim about what constitutes learning is, as we have seen in previous chapters, a matter of significant debate in the last decade of cognitive science. The idea that an agent completely decoupled from any immediate connection to the world can think about anything, let alone other minds, is questioned by enactivists and externalists in several compelling ways. Even where it might seem I am having a thought or experiencing an affective state that is wholly internal to me, if I am deploying certain concepts to think *about* whatever it is I am cognizing, that arguably means I am coupled to a sociolinguistic environment that extends well beyond my biological brain or body (cf. A. Clark 2003; Putnam 1975). Nevertheless, Penn and Povinelli assume that genuine learning is an internal process and that in order to think about the minds of others, one must be able to enter into this decoupled state. Nonhuman animals, it is argued, rarely if ever decouple from the immediate physical transactions in which they are enmeshed. This is why, according to the detractors surrounding nonhuman animal mindreading, what is *really* happening when it seems as though apes are attributing false beliefs to one another, or crows recognizing goal-directed action, is that they are relying on immediately present and visible actions and behaviors and drawing associations. What they are reading, in other words, are behaviors, not minds.

Another assumption Penn and Povinelli make is that thinking is representational in nature. This is of course not an assumption specific to them—indeed, the representationalist view of cognition remains fairly dominant in cognitive science and philosophy of mind.[5] Recall, for example, Bermúdez's argument regarding nonhuman mindreading and how it requires *intentional ascent*—that is, in order to think about the thoughts of others, one must be able to form propositional attitudes that represent beliefs about what the propositional attitudes in the others' minds are. Or, one must be able to think about thoughts, and this requires language, or what Bermúdez calls *semantic ascent*. Animals do not form propositional attitudes, he asserts, and thus are not capable of mindreading. Penn and Povinelli do not stress the propositional-attitudes component of representationalism as much as Bermúdez, but their line of reasoning is similar. They claim that other animals are indeed intelligent and complex beings who can extrapolate a great deal from their environments, including their conspecifics. However, they argue:

> Our principal disagreement is about the kind of representations over which these inferential and learning processes operate. The available evidence suggests that chimpanzees, corvids and all other non-human animals only form representations and reason about *observable* features, relations and states of affairs from their own cognitive perspective. We know of no evidence that non-human animals are capable of representing or reasoning about *unobservable* features, relations, causes or states of affairs or of construing information from the cognitive perspective of another agent. Thus, positing an f_{ToM} ... is simply unwarranted by the available evidence. (2007b, 737)

Here we can see both assumptions operating at once: the first being the claim that nonhuman animals only reason about objects or events that are directly present to them, and the second that thought is representational, while thoughts about others' thoughts require a sort of "metarepresentation." Animals can certainly form representations of the states of affairs of their world, but only from their own perspective. They cannot take up the perspective of another, because that would require reasoning about unobservable events and forming representations that are decoupled from their specific location in time and space.

I have more to say about these assumptions in the next chapter, and why I think they are problematic, not just for explaining mindreading, but also for developing a general theory of cognition that avoids the pitfalls of uncritical anthropomorphism. For now, I want to problematize the supposedly uncontroversial idea put forth by Penn and Povinelli that there is a clear and sharp line dividing what humans do when they observe others' intentional actions and what nonhuman animals do, and that this line is, as Bermúdez argues, a linguistic one. I am not convinced that only humans are capable of forming meta-representations about the propositional content of the minds of others. *Even if* this assumption turns out to be true—and I agree with Penn and Povinelli that it is an empirical question that needs to be tested—I do not think that the corollary, that this is the *only* measure of mindreading, follows. In other words, rather than arguing against what is an open and unsettled empirical claim, I want to challenge the inference that once this issue is settled, it will also settle the issue of whether other animals mindread or not. It is unwarranted, I suggest, to move from the claim "there is no evidence that nonhuman animals attribute propositional attitudes to other animals" to "no nonhuman animals can mindread." Moreover, I see a major problem with the further leap to the conclusion that in all cases except for humans, what we are observing is behavior-reading and not genuine mindreading.

Perhaps the most glaring if not implicit assumption lurking in many arguments surrounding mindreading is the all-or-nothing stance taken toward the capacity. Indeed, treating mindreading as a univocal capacity is itself problematic, as it assumes an ability or set of abilities that can be readily measured. If other animals that are incredibly distinct from humans—cephalopods or fish, for instance—are to have a fair chance at ticking off the requisite boxes for mindreading, it seems unfair to start the process by appealing to human-specific behaviors that are taken to indicate the capacity in humans. Andrews brings up this issue in her book *Do Apes Read Minds?* (2012) by noting that most of the folk psychology surrounding mindreading is premised on the idea that when we attribute mental states to others, we are attributing propositional attitudes to them. I might believe that "Mary believes that the toy is in the bucket," and I might also believe that "Mary doesn't know that the toy was moved from the bucket, but I saw that happen, so her belief is different from my own." While it is certainly true that we often attribute beliefs in this way, this is not the only way to understand and predict behaviors in others. For example, I might recognize that Mary is sad without attributing any propositional attitudes to her. As we saw in chapter 2, it is quite probable that a lot of affective attribution happens in this non-propositional way—I just *see* sadness in Mary's actions, facial expressions, body language. And my recognizing sadness in others allows me to effectively predict what they might do or not do. While this example is not necessarily me attributing *beliefs* to Mary, it is nonetheless a case in which I am mindreading, assuming that by "mindreading" we mean to include a wide range of skills and abilities to recognize what another being is thinking, feeling, planning to do, and so forth.

Based on what Andrews is arguing, I think we need to approach the question of whether nonhuman animals "read minds" in a more pluralistic and nuanced manner. First, as she suggests, we need to focus more on how humans "read people," not just "minds." Second, I don't think the answer will be a simple yes or no, the way many researchers seem to claim. Consider again the findings from Udell et al. regarding the differences among wolves, companion dogs, and shelter dogs in a variety of situations aimed to test their "mindreading" skills. Companion dogs were significantly better than wolves or shelter dogs at detecting that a human was less likely to give treats when pretending to read a book), but even companion dogs did not pass all the tests, such as the bucket-over-head scenario. In a response to these findings, Horowitz argues that for the test to provide convincing evidence of mindreading, subjects should pass in every situation and on every trial. The mixed performance of canids in this study, and of other animals in many other studies, leads researchers to conclude that in fact no other animals besides humans are capable of mindreading. However, as Horowitz further argues in the case of dogs, there is no definitive proof that they do or do not have a theory of mind, and "using the experimental paradigms developed thus far, there is no chance that any research could legitimately make such a claim" (2011, 315). There are probably experimental design flaws and reasonable explanations for the mixed performance. For example, some dogs may have been habituated to book-reading behavior and how it correlates with lack of treat dispensing by humans, but not be accustomed to bucket-wearing. What Horowitz suggests we consider instead is that dogs possess a "rudimentary theory of mind." I tend to agree with her, though I will further differentiate among levels of mindreading abilities. First, let's think about what a rudimentary theory of mind might be and why there is reason to posit its existence.

In one study, Horowitz (2009) found that dogs will use attentional cues to determine what type of attention-seeking behavior to deploy when requesting play interactions with humans. Humans who are most inattentive will elicit more forceful attention-seeking behavior. Horowitz suggests that findings such as these indicate that dogs are sensitive to unseen features that drive behaviors in others, and in this case the folk psychological term we use to describe the type of unseen vehicle of interaction that is implicated in many human cognizing tasks is *attention*. Indeed, as the previous chapters have argued, this feature is highly important in parent-child interactions, especially those that precede the onset of linguistic verbalization in the child. Furthermore, attention is a legitimate "unseen" vehicle of behavior that humans and nonhumans alike utilize, despite the claim of researchers like Penn and Povinelli that nonhuman animals never reason about anything other than what is immediately "present." These attention-involving interactions arguably lay the foundation for more robust mindreading in the child's future. Thus we come full circle to the beginning of this chapter in which we discussed the arguments surrounding mindreading in humans—what exactly it is, when it starts, and how it might be realized in the brain. Without delving back into that dispute, it is worth recalling that whatever theory one adopts to explain it, no one denies that neuro-typical, pre-linguistic children have the potential to become full-blown mindreaders. There are debates about when these capacities totally come online and what mechanisms subtend this process, of course, but *most* human children eventually develop a theory of mind. Arguably, therefore, the ability to recognize attentional cues is a step in that direction.

Not all humans eventually become proficient mindreaders, however. Those who fall somewhere on the Autism Spectrum Disorder (ASD) scale are often the first persons discussed in the ToM/mindreading literature, because a common feature of those with ASD is the inability to seamlessly utilize facial expressions, gestures, affectivity, or other social cues to properly decode what others are thinking. Simon Baron-Cohen (1997) has gone so far as to claim many people with ASD are "mindblind," and while I do not possess the expertise to weigh in on his assessment, I would note that just as ASD is a varied spectrum of neuro-atypicalities, the capacity to read minds

FIGURE 1. Original artist unknown. This image has become a meme on the internet to illustrate a variety of points, but it has also been utilized by researchers to discuss level-2 perspective-taking in humans, especially as this skill develops in children (see Surtees et al. 2011).

is arguably spectral, not an all-or-nothing skill, and is much more nuanced than it is often touted to be. Furthermore, this complex set of skills can be attenuated through specific interactions.[6]

Likewise in animals, perhaps, as Horowitz argues, many of them possess a "rudimentary" capacity for mindreading. Empirical evidence bolsters her claims, and not just in canid research. Evan Westra (2017) agrees that taking the perspective of another is not an all-or-nothing capacity, and he submits that it seems to occur on at least two levels. Level 1 is the type where I can know that if you have a blindfold on, you will not be able to see. Level 2, on the other hand, means I can not only take your perspective based on immediate physical or environmental factors such as being blindfolded, but I can recognize that what you are seeing or not seeing is the result of a particular *visual gestalt*, or perhaps even the product of an extended web of sociocultural practices. One example to illustrate level-2 perspective-taking is seen in figure 1. Here the person on the left sees a 6, while the person on the right sees a 9, but if the former can take a level-2 perspective of the latter, that person knows that the reason the other person sees 9 and not 6 is because of the other person's situatedness in the world.

I like this example because it illustrates how multifaceted and complex mindreading can be, and why it most likely admits of degrees. Again, cognitive ethological findings support this, as Westra notes that several animals—canids, corvids, and great apes among them—have been shown to engage in level-1 perspective-taking, and even though we have not found evidence that they engage in level-2 perspective-taking, there is reason to think that by engaging in the former, they are, in some basic sense, reading the minds of others. We can further add nuance to this account by utilizing the awareness-relations framework put forth by Martin and Santos. Level-1 perspective-taking would also include the ability to track another's awareness and use that information to plan and predict future actions, whereas level-2 would involve much more—false belief attribution, imputation of ignorance or knowledge to other agents, and the like. Much as it is an oversimplification to think of pre-linguistic children or persons with ASD as "mindblind," the same oversimplification is present when we classify all nonhuman animals as incapable of mindreading solely on the basis of how they perform on human-specific tests aimed at measuring a high-level form of what is assuredly a myriad of skills.

Instead, like Westra, Horowitz, and other researchers including Butterfill and Apperly (2013), I think mindreading should be placed on a spectrum, and I have developed my own version, which is even more detailed than most, simply because I want to establish a better means for explaining the capacities that nonhuman animals—and especially dogs—exhibit. In the case of dogs, as I have argued so far, some of these capacities seem to be uniquely tied to the interactions the dogs engage in with humans. For example, companion dogs in the study conducted by Udell et al. were likely only able to determine the proper behaviors to engage in because their umwelt was already imbued with these human artifacts and with humans who might appear distracted or engaged with them on a daily basis. Or, in Horowitz's study, we might say that the dogs who use more forceful play elicitations have learned that this behavior works better than more moderate approaches because the distracted human is a regular part of their lifeworld. Not only can a mindreading spectrum address these nuanced differences, it can also get us closer to answering the important question cognitive ethology demands. Besides knowing *how* something might occur in an animal, we need to address *why* it does. Thus the spectrum I have worked out is as seen in table 1.

Notice that I refer to the middle levels—1 and 2—as forms of "social cognition" rather than calling them "mindreading" or "theory of mind." This is because I think, as Andrews, Horowitz, and others have argued, that a lot of what humans do when they are engaging in interactions that help them understand others does not fit the standard folk psychological framework of

TABLE 1. The Spectrum of Social Cognition Capabilities

Level	0—Mindblind	1—Fundamental Social Cognition (FSC)	2—Second Order Social Cognition (SSC)	3—Complex Theory of Mind (CTM)
Description	Unable to recognize other minds; unaware that other beings possess intentionality. Might see actions as goal-oriented, but not stemming from thoughtful planning	Tacitly aware that other beings possess thoughts and intentionality; able to perceive others' actions as goal-oriented and the result of thoughtful planning; may recognize that other perspectives are different from one's own; possesses and utilizes awareness relations to predict and act	Explicitly aware that other beings possess thoughts and intentionality; perceives goal-directedness stemming from thoughtful planning; recognizes that thoughts of others are not the same as one's own, due to difference in perspective or belief; can take up the perspective of others at least briefly; utilizes ignorance-knowledge conditions to predict and act	Recognizes not only that others might have different thoughts and beliefs, but that these differences might be the result of long-standing perspective-taking, socially construed belief systems, cultural norms, etc. Can take up the perspective of another and empathize over temporally extended periods

attribute X propositional attitude to subject A during this or that action. Sometimes recognizing what another being is thinking, feeling, planning, or attempting amounts to here-and-now transactions where outward behavior, along with unseen features such as attention, *just are* parts of the social cognizing that is going on. Furthermore, as Barrett et al. convincingly argue, it is anthropocentric to use high-level theory of mind as the sole litmus test for social cognition. Nevertheless, when nonhuman primates demonstrate social complexity, they argue, it is equally tempting to assume cognitive complexity underlying such behavior. To reconcile this tension, they suggest instead that "apparent cognitive complexity may emerge from the interaction of brain, body, and world" and, further, that "under these conditions, individuals do not need to hold abstract conceptual notions of 'bonds' or track others' relationships because they can gauge circumstances directly by looking at what is happening around them: the spatial structuring of the animals in their environment may obviate the need for certain kinds of high-level processing in the animals themselves" (2007, 568). Echoing philosophers like Andy Clark (1993, 1998) and scientists like Brooks (1999) and Johnson (2001), Barrett et al. go on to suggest that these dynamic social interactions do not indicate some underlying cognitive mechanism, but instead the interactions *just are* the cognitive processes.

Although Barrett et al. are discussing nonhuman primates, I think their account applies to many forms of human social cognition whereby apparent social complexity need not depend on complex, internal, representational mechanisms. Likewise, if other animals show similar social complexity in interactions, I think their account would help bridge that explanatory gap equally well. Dogs, as we have seen, are quite skilled at fundamental social cognition (FSC). And while I agree with how Martin and Santos (2016) interpret the findings pertaining to nonhuman primates—namely, that they fail to represent ignorance-knowledge in others—I don't think this rules out entirely that nonhuman animals can possess elements of second-order social cognition (SSC). As my chart points out, and as I have discussed throughout the book, it is not altogether settled (1) the extent to which thought is representational and (2) just how much cognizing can be entirely "decoupled" from the environment. Martin and Santos assert that decoupling is what is required for an organism to think about the thoughts of others and represent them as "knowledgeable of" or "ignorant to," but as I have argued thus far, enactivists like Hutto are not convinced of this and have made compelling cases to the contrary. While I don't think I can settle that debate once and for all here, I have noted in the chart that in order to possess second-order social cognition, an organism must *utilize* ignorance-knowledge conditions in purposeful action or in predicting/planning. Characterizing it this way leaves open to what extent the organism is *representing* that information. As I argued in chapter 2, using information meaningfully does not require an internal, computational model of the world. Indeed, some of the research, especially Hare's studies regarding how dogs tend to use human attention to determine when to obey, not obey, sneak food, and so forth, suggest that there are elements of SSC at work in their cognizing, whether they represent the human qua "ignorant" or not. The ultimate utility of social cognition—no matter where on the spectrum—is providing an organism with means to communicate, plan, predict, and purposefully interact with other social organisms. Though other animals besides humans might not have heretofore demonstrated complex theory-of-mind (CTM) capacities, it makes little sense to rule them out from possessing any sort of "mindreading" skills, since, as we have seen, the skills are so varied and spectral.

Conclusion: Perhaps There Is More to Mindreading?

All of this discussion about how dogs might read minds, how they might extract social and emotional information from us and from one another, and how clever they can be has centered experimental paradigms that measure dogs' capacities in fairly structured environments. The dogs are asked to solve problems, figure out what a human is communicating, or learn when is the best time to beg for food. One place we have not yet looked in order to examine how dogs think "with us" is in how they play with us. Play is such an obvious source of interaction between humans and dogs that it perhaps goes unnoticed because of its ubiquity. The difficulty of setting up effective experimentation might also prove to be a consideration. There are, however, more and more findings emerging from studies of human-dog dyadic play, and all those findings, I will argue, point to a similar conclusion: that the continuum of social cognition applies quite well in explaining the behaviors observed. Accordingly, the next chapter is devoted to play, not just as a recreational and fun interactive exchange we have with dogs, but as a genuine form of thoughtful action and coactive cognition between two otherwise distinct species. Play, in other words, is quintessential coactive cognition.

CHAPTER 4

Thinking-in-Playing

SOCIAL COGNITION BEYOND MINDREADING

An Overlooked Component to Social Cognition: Play

Chapter 3 argued that mindreading is not a univocal capacity, nor is it an all-or-nothing skill. Rather, it is best thought of as occurring on a spectrum, and this spectral characterization of mindreading applies to the ways humans and nonhumans mindread among one other, as well as to interspecies social cognition. In human-dog dyads there is compelling evidence that interspecies mindreading is taking place, as we have seen exemplified in studies conducted by Hare et al. and others.

I also argued that examining interspecies social cognition further buttresses the idea that the Theory Theory/Simulation Theory dilemma is an outdated and false dichotomy, supporting instead a third option, an account that emphasizes primary intersubjectivity. Often referred to as Interaction Theory, this approach not only provides a better account of human-human mindreading but is best suited to explaining how there can be meaningful exchanges between humans and dogs that do not rely on theory formation or simulation, per se. Though those elements might indeed figure in at some points, what I intend when I am interacting with my dog is directly perceived in my gestures or facial expressions, much in the way that my dog's wagging tail and excited jumping as I walk in the door after a day at the office indicate happiness to see me. And again, this is not because I infer these behaviors as analogous to my own, nor because I theorize that these behaviors typically indicate this or that emotion.

Finally, I attempted to draw all of this together with the enactivist account of cognition I have been supporting thus far in explicating how we think-with dogs and they with us. But I don't think that story has been completely and convincingly told yet. In this chapter I want to explore further the ways in which the interactions are themselves part of social cognition specifically and of cognition more generally as these processes occur when humans and dogs engage with one another. In a recent paper, Gallagher suggests that the very capacity for understanding another person (or dog, for that matter) is arguably a "social affordance" that relates to and depends on "possibilities opened up by interaction itself" (2020, 63), and I think this is an apt jumping-off point for this chapter. In what ways can it be argued that interspecies communication and meaning are afforded by specific modes of social transactions? I have hinted at answers to this question already, but it is in this chapter that I think a more concrete answer might be given.

At the end of chapter 3, I suggested *play* as an overlooked type of interaction between humans and dogs in which we might find genuine social cognition happening, and moreover the sort of cognizing that is, once again, best captured by an enactivist picture. In this chapter, therefore, I want to *take play seriously* and examine more closely the ways in which social cognition emerges from and is sustained by human-dog encounters. Play, as it turns out, provides perhaps the richest data set for demonstrating that not only do humans and dogs recognize intentionality and goal-directedness in each other, but this awareness is both a prerequisite for and a product of those meaningful interactive dyads.

I begin by providing a general overview of play, as it occurs between humans, between animals, and between humans and animals. In doing so, I keep an eye to the way play has been argued extensively—both in developmental psychology and in cognitive ethology—to facilitate development, cognitively, affectively, and socially. Next I revisit a worry that I've raised throughout the book: anthropomorphism. It is easy to get carried away when thinking about how dogs and humans play together and to impute all sorts of what might arguably be human-specific capacities to the dogs, with little or no evidence to support these attributions. Indeed, if a dog performs a perfect run through an agility course, we might characterize the dog as having known exactly what the human wanted or as being adept at reading the intentions of the human in the interaction. It could be, however, that dogs are just good at associating behavioral cues with actions and that they don't necessarily know that a human intends anything whatsoever. I think this concern can be ameliorated by first appealing to the continuum I proposed in chapter 3, but also by reviewing some of the empirical work on play between humans and dogs. Most notably, I focus on studies conducted by Robert Mitchell and colleagues, as they lend support to the idea that a sort of rudimentary mindreading is taking place, what might be referred to as a "conversation of gestures," as George Herbert Mead called it (1934, 42–43). Not only will play provide a compelling case for the sort of second-order social cognition that I suggested in chapter 3 at least some dogs possess, but the analyses of play we will encounter successfully resist any uncritical anthropomorphism.

I then turn to discussions of animal studies from critical theory, especially those found in Donna Haraway's and Vinciane Despret's perspectives, to further argue against the potential claim that my account is overly anthropomorphic or, worse, anthropocentric. Through this discussion, we come upon the notion of a *sympoietic system*, which is a system comprising multiple organisms, tools, and/or social structures. Sympoeisis, or *making-with*, serves as a useful analogue to how humans and dogs play, thereby constructing new forms of thoughtful action. This idea can be further bolstered by the *holobiont* theory in biology. I use these ideas and frameworks to think about how, especially in play, the interactions between humans and dogs create and sustain a variety of coactive cognition, rather analogous to what scientists often argue is the purpose of play in human children.

Finally, I address a concern regarding the enactivist account that I am defending. Enactivism often rests on thinking about organisms as *autopoietic*, as we have seen. In other words, they are self-creating, self-organizing, and self-sustaining systems, which are at once stable and predictable but also dynamically open and constantly changing. This would appear to be at odds with my reliance on sympoiesis and reframing human-dog play dyads as systems that "make-with." Despret (2008) argues that in many cases of human-animal interaction, the very meaning of subjectivity is itself altered such that what arguably emerges is not a separate human playing with a distinct dog but rather a human-with-dog. This is precisely where I argue sympoiesis becomes a useful tool for describing the transaction, but it also seems at odds with some basic tenets of enactivism. Thus I end the chapter with an explication of this tension, and suggest how we might resolve it.

How Humans and Other Animals Play

Play occupies the pages of countless articles and books authored by psychologists, linguists, and educators. While I will not delve too deeply into the many multidisciplinary approaches to and reasons for studying play, it is worth noting that while numerous disciplines have continued to emphasize the crucial role of play in the development of such skills as language and pro-social attitudes, philosophers have remained relatively silent about play. This is not to say that disquisitions on the subject are nonexistent, but for the most part, what philosophers mean by "play" is often quite different from what psychologists mean. Wittgenstein ([1953] 2009) famously problematized play by asking how best to ontologically classify a game, and ever since then, philosophers have been grappling with defining and cataloging what constitutes play, games, sport, and the like. These are important issues, and I do not mean in any way to diminish the metaphysical work being done by thinkers taking on the task, but for our purposes, it's not important to develop a precising definition of a concept whose denotation is wide ranging and probably vague or "fuzzy." Indeed, as I argued regarding what constitutes mindreading, I contend that play also occurs on a spectrum and to focus too much on the game-theoretical aspects of it, for instance, is likely going to cause us to overlook the ways in which nonhuman animals might engage in play as well.

It is interesting to note that Plato had quite a bit to say regarding play, and much of it mirrors what psychologists today tend to think about the function of play. In the *Laws*, book 1, he says: "When children are brought together, they discover more or less spontaneously the games which come naturally to them at that age" (1997, 794a). Plato urged, however, that this "free play" will actually help satisfy certain educational goals—kids will learn important features of farming, trading, manufacturing, and so forth, while playing with others. In the *Republic* he says, "Do not keep children to their studies by compulsion but by play" (536e–537a), further emphasizing that it is play that will allow children to achieve educational ends, not forced study. This "radical" notion that play is necessary for innovation and development of ideas would not be resurrected until hundreds of years later, and it is now seen in debates about school recess, inclusion of the arts in public education, and similar issues, where many argue, like Plato, that play is just as important for learning as formal instruction, if not more so. Part of why Plato's ideas took so long to resurface was likely Aristotle's placing play squarely in opposition to work. "Leisure," Aristotle argued, was what allowed for educational insights, not play. Play was simply, according to book 8 of his *Politics*, a break from work: "We should ask what activity real leisure consists of. It's certainly not playing. That would mean play was the be-all and end-all of life, which is out of the question. The fact is that play relates to work more than to leisure: the worker needs a break, and play is about taking a break from work, while leisure is the antithesis of work and exertion" (1946, 1337b). The true mark of the educated person, therefore, was plenty of leisure time in which to muse and mull over what had been learned and to generate new ideas. Play was something little kids engaged in, but it was not serious, and certainly not educational.

Fast-forward to current philosophical discussions of play and you will be hard pressed to find much about the educational benefits of play. Ryall et al. (2013) examines philosophical perspectives on play, and the essays in it are among the few to break away from the metaphysics of play that seem to have dominated the landscape ever since Wittgenstein, but there is very little in the way of the pragmatics of play. Perhaps the closest we come to seeing Plato's ideas exemplified is in the context of sport-as-play. I will return to sports and what some philosophers have to say about them regarding play later in the chapter to draw some analogies to the overarching arguments I'm making in this book, but for now, suffice it to say that I shall not be drawing much from philosophy in my discussion of play that follows. Instead I turn to psychology, where play between humans has been extensively studied and, for the most part, lauded for its developmental benefits.

Human Play

Psychologists have been demonstrating the benefits of play in human development for quite some time now (see Broadhead et al. 2010; Moyles 2014; Whitebread 2017). When children play, they are most often unwittingly participating in a sort of educational exchange that allows them to form new associations regarding social rules such as turn-taking, cognitive skills such as spatial reasoning, and emotional transactions such as facial expressions and gestures. Likewise, Vygotsky extensively argued that "pretense play" is instrumental to language acquisition, and neo-Vygotskians continue to demonstrate the connection between learning to "play pretend" and mastering important linguistic skills (see Vygotsky 1986; Winsler and Naglieri 2003; Fernyhough and Fradley 2005).

Regarding emotions, as we saw in chapter 2, a lot of affective intelligence comes online precisely through embodied interactions that help to co-regulate and co-attune the emotional experience. Play, it turns out, helps this process along. Berk et al. (2006) note the role of pretense play in the development of emotional self-regulation. Several studies examine how children learn to cope with emotionally arousing or stressful events, particularly by utilizing forms of pretense to act out those feelings. Children will often engage in socio-dramatic play relating to stressful or traumatic situations arising in their experience, such as getting a shot at the doctor's or getting bitten by a peer at daycare. Incidentally, a whole genre of "play therapy," based largely on Carl Jung's psychological theories, has come to be a popular and effective form of treatment for children as well as adults who suffer emotionally from abuse, grief, or other negatively impactful conditions (see C. Clark 2006).

Notice that in the discussion so far, I have not provided a working definition of play, but rather focused on the pragmatics of it in terms of development. As noted, there are some really intriguing discussions pertaining to play ontology and whether a unifying definition is even possible. But, I maintain, that deeply philosophical issue about the proper definition of play need not concern us, for a couple of reasons. One, we can observe play-like behavior in humans and nonhumans and be pretty sure we are indeed witnessing play without having such a precise definition. To be sure, we are likely to overlook some important aspects of what play might consist in, and we likewise run the risk of over-ascribing play to situations that are in fact not play. However, given the scope of this project, I am only focusing on some pretty well-documented and agreed-upon instances of play, so that risk is nearly nonexistent. The other reason I don't think a precise and unifying definition of play is needed is that a taxonomy of play behaviors will prove much more useful. As with mindreading, I think play occurs on a spectrum of possible iterations. This idea is not my own. In fact, for several decades now the dominant way of categorizing play has been to arrange it according to types. Interestingly, those types are defined in terms of their role in development. In other words, we can see Plato's radical idea that play is a necessary component of learning borne out in this taxonomy, which itself mirrors basic development. Here are the categories, briefly.

PHYSICAL PLAY. Physical play includes jumping, running, dancing, skipping, rough-and-tumble play, and fine motor play, such as coloring or drawing. Evidence suggests that this type of play helps children develop hand-eye coordination and proprioceptive awareness and is crucial for basic strength and endurance (Pellegrini and Smith 1998). Rough-and-tumble play, or what is colloquially called roughhousing, is the most extensively studied type of physical play. Research indicates a clear association between roughhousing and the acquisition of affective skills and social cognition. It is associated with forming strong emotional bonds, trust, and what might be collectively termed "emotional intelligence" (Jarvis 2010).

OBJECT PLAY. Widely observed in all primates (see Power 1999), object play includes teething, mouthing, and touching objects, arranging objects, rotating objects. This type of play is often referred to as "sensorimotor play," and it helps with physical problem-solving skills (Pellegrini and Gustafson 2005), facilitates private speech and self-commentary, and aids in the development of perseverance and a positive attitude toward challenge (Sylva et al. 1976).

SYMBOLIC PLAY. Symbolic play includes babbling, repeating sounds, repeating words, making up new words, rhyming, and eventually recognizing and even making jokes with and about language. There are obvious benefits to this type of play in terms of linguistic development. Symbolic play also includes musical play. Trevarthen (1999) has shown that playing with musical rhythms facilitates recognition of patterns in speech. And Kirschner and Tomasello (2010) performed a study involving four-year-olds where they found that children involved in joint music-making play showed significantly higher levels of cooperative behavior than the control group of children who did not participate in any such play.

PRETENSE PLAY. Pretense play includes all sorts of "pretend play," such as pretending to use a phone, or using a nonphone object (like a banana) as a phone, imaginative play with toys, and socio-dramatic play, such as dress-up or playing "house" or "doctor." Vygotsky was the first to extensively study this sort of play and noted its role in developing linguistic capacities such as metaphor and complex sentence structures. Karpov (2005) further argues that dramatic play drives the development of private speech, which in turn allows for complex linguistic constructions in outer speech. Berk et al. (2006) also note that this form of play is important for learning self-restraint and self-regulation, which both aid in the development of pro-social skills.

RULE-BASED (GAME) PLAY. Rule-based play includes any game with specified constraints, rules, or instructions, such as chess, Go Fish, or Marco Polo. Adults, of course, engage in this type of play quite often. For young children, rule-based play helps to learn a range of social skills related to sharing, taking turns, and understanding others' perspectives (Fromberg and Bergen 2006).[1]

In short, play, according to many, is not ancillary to but *necessary* for healthy development in humans. This is of course not to say that all forms of every category of play must be enacted for a person to fully develop. But play, in some form or another, sets the stage for crucial developmental milestones, such as mindreading, and it is perhaps unsurprising that human children who show certain deficits—those with ASD, for instance, who might not understand metaphor—did not or do not engage in symbolic or pretense play. Thus, as theorists such as Trevarthen (1999) argue, the stage-setting nature of play is important for a large array of cognitive and social skills, from perspective-taking to affective regulation. As we have seen, in the nonhuman animal world, components of social cognition such as turn taking, co-attunement, and mindreading arguably exist, and they are exemplified in dogs' interactions with humans. Much like human children, dogs are quite keen on play, as are many nonhuman animals, and it is not a big leap to argue that the same connection between acquiring cognitive and social skills and play exists in the nonhuman animal world. Let us, then, turn our attention to them, focusing particularly on dogs, so we can examine this connection a bit more closely.

Nonhuman Animal Play

Nonhuman animals have been widely observed to engage in a variety of play types. Although mammals are the most extensively studied, play behaviors are certainly not limited to this class. Octopi have been observed to explore Legos that are placed in their tanks, thereby exhibiting a sort of object play (Kuba et al. 2006). Reptiles, such as soft-shelled turtles, apparently play with objects inside their shells (Burghardt et al. 1996). And all sorts of birds have been seen exhibiting play behavior, among them herring gulls, who will "play with their food," so to speak, by engaging in drop-catch behavior with crabs or other shellfish, rather than eating them (Gamble and Cristol 2002). Despite all these intriguing findings, the fact remains that play is primarily observed in mammals (e.g., Resende and Ottoni 2002), and it is primarily among mammals that other forms of play, such as social play, are seen.[2]

As Bekoff (1984) aptly notes, play is one of those things that is not easily defined but is immediately identifiable when it occurs. Part of this stems from the fact that play is quite *context-sensitive*. The same action—pushing/shoving, say—could be considered play or violent aggression, depending on what else is going on. So, first, when attempting to isolate and observe play behaviors in any species, we are already in a critical position, knowing all these complexities surrounding play and how difficult it is to precisely define without actually being involved in the observation of play itself. In keeping with the critically anthropomorphic methodology of this book, I think utilizing the criteria found in Gordon Burghardt's book *The Genesis of Animal Play: Testing the Limits* (2005) will be especially useful. Burghardt outlines five requirements that must be met if we are to consider the animal "playing":

1. The activity must not be fully functional. That is, it must not be a necessary component of the animal's survival in the particular context in which it occurs.
2. It must be spontaneous, voluntary, intentional, pleasurable, rewarding, or autotelic (done for its own sake rather than as instrumental for some other end).
3. It must be different from any other ethotypic behavior, structurally or temporally. If it is to count as play, in other words, it will "look" different from the other activities of the animal.
4. It should be repetitive but not stereotyped. That is, play patterns should be observed, but if these behaviors are more like stress-induced "tics," such as obsessive grooming, then they are not playful actions.
5. These activities must occur when the animal is free from danger, in a relaxed state, not hungry or thirsty.

Burghardt's criteria allow for all sorts of animals to be considered "playing" in varying contexts, and because he relies on a framework for play that is sensitive to the particular umwelten of the animals, he avoids being guilty of anthropomorphism by omission. It is worth noting at this point that anthropomorphism is unavoidable if we seek to ascribe "play" to any nonhuman, as the word itself is inextricably imbued with human significance. However, Burghardt's criteria arguably strip play of its anthropocentrism as much as possible, leaving us with a set of requirements for an action to be considered playful that are as close to objective as we could hope to get. It is indeed not possible to remove ourselves from the equation here—we are the linguistic progenitors of the term "play," and as such we get to stipulate what counts and does not count. But because Burghardt takes care to define play as inherently dependent on the particularities of the animal's lifeworld, I think we have an excellent case of critical anthropomorphism to work with, and one that will allow for genuine scientific and philosophical inquiry.

While many nonmammals arguably play, let us briefly focus on the mammal in question in this book, *Canis familiaris*, to see how Burghardt's criteria provide good reason to believe what most of us already intuit: that dogs are quintessential "players." I often observe my two dogs, Darwin and Tesla, engaging in a bout of what I consider to be play—they have a rope, or some kind of "dog toy,"[3] and are tugging at it in what humans call a tug-of-war. One of them will inevitably lose grip and the toy captor will dart away, toy in mouth. The other dog will give chase, and the toy captor will keep the toy just out of range so the other cannot grab hold. This goes on for a while, and then finally the toy captor allows the other dog to get hold of the toy so the process can begin all over again. Are my dogs playing? What they are doing is certainly not functionally necessary for survival at this moment. They are safely nestled in the warmth of my living room, where the most dangerous predator is a three-year-old toddler who might yank a tail if he is unsupervised. This type of activity often occurs spontaneously. It is not clear whether it is rewarding or pleasurable, but arguably the dogs meet criterion number 2, at least to some extent, as their activity seems obviously *autotelic*. This activity is not just an ethotypic behavior of the dogs. Although this "game" I see them engage in so often is *typical*, it is not just another "dog behavior." It is temporally structured—one dog takes the toy away and disallows the other to get it for some time and then allows the tug-of-war routine to continue, and so forth. In this way, the dogs' activity is repetitive but not stereotypical. This "game" occurs frequently in my house, and it is generally similar each time, though there are some interesting "improvisations" I observe. And last, the dogs almost never engage in this activity if it is close to dinnertime, if their water bowls are empty, or even if we are out at the dog park where there are distractions, such as other dogs, that could be perceived as threatening. In short, it is overwhelmingly clear to me that Darwin and Tesla are playing here.[4]

We could contrast this example with instances where the dogs are seemingly playing the same game but are doing so when, for instance, I have human company over. Especially if the persons are unfamiliar to my dogs, it might be less convincing that we ought to call their behavior play, even if it looks just like the bout I described above. They could be reacting to a potential threat from the newcomers or just overly nervous and excited. I often joke that they are "showing off," which is a noncritically anthropomorphic way, perhaps, of utilizing Burghardt's second and fifth criteria to argue that they aren't really playing in this case. They are trying to *get something*—attention, maybe?—or they are stressed and reacting to that stressor. Of course, they could just be playing, and that interpretation might win out in the end, but the point is that in these two different scenarios, the former is far less controversially an example of play, while the latter would need to be investigated more.

At least some nonhuman animals play, dogs included. This is a fairly uncontroversial claim. But *why* do they play? Comprehensive cognitive ethology would demand that we address this question. There are many reasons that might be postulated as to why mammals are more likely to be seen playing—longer maturation period, neurophysiological differences, greater social complexity—but those speculations are not relevant here. Regardless of whether we are examining mammalian or reptilian play, a common thread runs through most ethological analyses of play and when it tends to occur. The general hypothesis is that play occurs when certain environmental and physiological conditions are met—no predators threatening the animal, lack of hazardous climate conditions, the animal in a relaxed state, and so forth. The herring gulls mentioned above were observed engaging in drop-catch play behavior only when the weather was sufficiently warm, and the octopi would not play with objects in their tanks if food-deprived. When we consider companion dogs—at least those dogs who live in homes with humans and enjoy being full-fledged members of a human family—it makes perfect sense that a lot of playtime would be afforded, given the relatively low levels of stress and hazards these dogs face daily. I discuss canine play in detail below, as it is the focus of this chapter, not just because dogs, like apes and other mammals, play a lot, but because dogs are arguably the species that plays with humans the most, aside from other humans themselves.

But we are still talking about *how* play can occur and not really addressing *why* it occurs, or more appropriately, what purpose it might serve, both for the individual and for the species. This is of course different from Burghardt's argument that if the activity is to count as play, it must not be functional. This criterion applies to specific bouts of play and claims they cannot be necessary for survival. In asking what purpose play in nonhuman animals might serve, we are instead asking about the development of the animal and the evolution of its species. We know that human and nonhuman play are categorized similarly. Ethologists often divide play among *locomotor*, *object*, and *social* types (Oliveira et al. 2010). This categorization is similar in some respects to how psychologists think of human *physical*, *object*, and *pretense* play. Likewise, as is the case with humans, ethologists have pointed out how crucial play is for so many of the important cognitive, affective, and social skills the animals need to acquire, not only to survive, but to thrive. For example, by learning from experiences in various social situations, animals can improve their capacities to compete with members of their own species, as well as with members of other species. Play is also known to build group cohesion in social contexts (Bekoff and Byers 1998). Findings such as these provide animal welfare advocates with persuasive fodder, such as the claim that dogs ought to be played with regularly, not left alone in an apartment all day.

When it comes to play in canines, scientists have been studying this behavior for quite a while. Unlike the reluctance to take canine intelligence seriously due to domestication, play has been viewed as an activity at which dogs excel, and it might just be *because of* their bonds with humans. While researchers have been examining play behavior in canids for decades, it has been the work of ethologists such as Bekoff, and psychologists such as Mitchell and colleagues, that has emphasized play as integral to the rich cognitive and social lives dogs lead. Bekoff is of course well known for work focusing on a variety of animal species, and has even argued that nonhuman animals can have a sense of justice and morality (Bekoff and Pierce 2009). He has also extensively studied the play bow in dogs, arguing that it signals an intent of nonaggression or clarifies the meaning of an interaction among dogs. Mitchell and his collaborators have since undertaken even more extensive research into canine play, and have extended this to include not only *intraspecies* but *interspecies* play as well, which occurs most notably between dogs and humans.[5] I discuss the work of Mitchell et al below in a section devoted to human-canine play.

In sum, nonhuman animals—dogs in particular—assuredly play. They engage in physical, rough-and-tumble play, and there is good evidence that many mammals engage in object play as well. While social play is also observed, it remains a matter of debate precisely what types of social play are occurring. For example, do chimps really engage in pretense play? If so, does this mean that they are reading each others' minds? We have already had this discussion in chapter 3, but it is interesting to reconsider it now in light of play behaviors in which they have been observed to engage. The main takeaway so far is that play, for humans and nonhuman animals, is essential for development in many ways. It is important to keep in mind, however, that while play serves this crucial purpose, it is still not the same as learning. It is not, in other words, "work." Play is fun. It is freeing. In his 2009 book *The Role of Play in Human Development*, Anthony Pellegrini argues that in play, as opposed to work, the participants can focus on the activity itself. Instead of having a specific goal to achieve or some task that must be performed to finish a project, play *just is* the project. In play, humans and animals can try out new behaviors, modify behavioral patterns based on dynamic feedback, and change the sequence of behaviors, all as part of an unfolding process that is almost always unbeknownst to the individual players, helping them learn more about their physical, social, and emotional world. Mitchell (2015) and Mitchell and Thompson (1986b, 1990) refer to this aspect of play as being a center-focused rather than an end-focused activity. In other words, the focus is precisely on the activity itself and is not really for any specified purpose. And yet, from play interactions, it is often the case that new skills emerge or even that meaning is generated. I have been arguing throughout this book that we can see this sort of dynamic emergence of thought, emotion, and social cognition best when we look at the human-canine dyad. Let us now turn our attention to this unique relationship, as I examine more of Mitchell's and others' important studies on dogs and humans working "at play."

Interspecies Play: Human-Canine Dyads and Coactive Projects

One way to think of what play generally involves is to consider how "projects" function in playful activity. Much like the concept of play, "project" can be a tricky notion to define, but for children or nonhuman animals engaged in play, it need not be a super-precise term. Mitchell uses Simpson's 1976 conception of projects in order to examine how these emerge in play in dogs and between humans and dogs. For Simpson, a project is just a pattern of actions that helps control the situation or the other participant(s) in the "game." Projects often involve "experiments" utilized to gauge what variations in behavior imply, how participants might gain or lose control over a situation, or how much variation in patterns or movement will be tolerated. And this experimentation can lend itself to the formation of "routines," such as would be seen in a game of fetch, where an object is thrown, retrieved, thrown again, and so on. Routines, of course, are not static, nor are they merely repetitive actions. They can change as variations are tested and new projects emerge. For example, if my dog Tesla is entreating me to play fetch by bringing me a toy to throw, and I throw it from the porch into the yard for her a couple times, she will often, after a few rounds of this routine, stop bringing the toy all the way back up to the porch for me to throw it again. Instead she waits at the bottom of the steps. Perhaps she is tired of going up and down the stairs, or perhaps she is bored with the routine generally. I do not profess to know the reason. But this variation in her behavior elicits a different response from me. Sometimes it means I simply acquiesce and descend the steps and begin playing fetch in the yard, which will often progress into a game of chase as Tesla slowly shifts into a keep-away pattern, or she may fetch a stick and entreat me to play tug-of-war. Other times, if I am feeling particularly lazy or mischievous, I will ignore her altered play behavior and pretend to be engrossed in something on my phone. I have learned that this is the surest way to get her to bring the toy all the way back up the stairs. If I choose this option, it will elicit a different pattern of behavior from Tesla, and a whole new series of routines will emerge.

Notice that in this short description of play and the routines and projects in which Tesla and I engage, there is never a final "point" of the play. Mitchell argues that this is because play is generally a center-focused rather than an end-focused activity. In other words, play is just about the playing. Tesla and I are not trying to accomplish a task necessarily, nor are we jointly attending to some common goal. Nevertheless, there are what we might call "micro-goals" at work. Or, to put it differently, my actions are goal-oriented, as are Tesla's. In a study reported on in 1986, Mitchell and Thompson argue that the findings indicate that dogs understand and can engage in deception, precisely because they can perceive actions as having this goal-directedness. In the study, human-dog dyads, the members of which had never met, developed projects for play that all involved deception, such as enticing or playing keep-away. Both humans and dogs would, in various ways, "con" one another—for instance, by a dog enticing a human to try snatching a ball from its mouth, or a human letting a dog get just close enough to an object and then snatching it away and holding it too high for the dog to reach. All of these cases, according to Mitchell and Thompson, were not only examples of how both dogs and humans use deception in play, but they indicate that dogs, just as much as humans, are aware that actions can be goal-directed. "That is, dogs and people can deceive because each can assess the projects of the other" (1986b, 200). This further implies, they argue, that dogs and humans recognize intentionality in the other's actions. In all the forms of play that emerged in this particular study, not only were dogs and humans able to assess the ongoing projects and goal-directedness of the various actions of their playmates, but they *used* those assessments to make predictions about future actions, which in turn guided their own actions so as to control, augment, or alter the play (202). As Dennett (1987) reminds us, if we get more explanatory power by taking up an *intentional stance* toward behavior (as opposed to a physical or design stance), then it is probably the correct way to explain the phenomenon. If, as Mitchell and Thompson observed, dogs consistently appear to alter their behavioral patterns based on the actions of the humans, and these alterations in turn afford stronger predictive abilities, then it is reasonable to conclude that the dog "sees" the actions as having intent behind them, as being *about something*, and as tending toward a goal. A dog might, for instance, retrieve a ball, bring it close to the human, drop it, and then allow the human to get *just close enough* before snatching it back up and running off, as a sort of "Ha! You really thought you could get it that time, didn't you?" The human in turn is compelled to feel deceived, and this is not egregious anthropomorphism on the human's part.[6]

It is important to note that just because someone perceives an action as goal-directed or as full of intentionality, this does not imply they see the *agent* of that action as a being that has intentionality. In another study, Mitchell (2015) cites Buytendijk (1936), who compares playing between dogs with humans playing sports. In both cases, Mitchell says, the players can deceive, but it is not "well-thought-out reason that directs the action, but an unconscious realization of the possible moves of the adversary" (2015, 34). As with sports, this ability to predict and plan based on the possible moves of the "adversary"—or simply the other players in whatever game exists between human and dog—can lend itself to creative moves or novel routines. The capacity to *create* and to invent within the established parameters of a game is not only easily demonstrated in human-dog play, but it is a hallmark of the sort of cognition I have been urging that we pay attention to throughout this book. In other words, thinking-in-playing, as it occurs between dogs and humans, is precisely the type of cognition that can only be properly understood within an enactive and intersubjective framework. Before I offer that argument, however, I want to pause briefly to consider a worry that has been a constant background concern, namely: In all the descriptions of play behavior and the explanations I've so far given for them, could it not be that I have slipped into an uncritical anthropomorphism? Could it be, for example, that in play, dogs are not actually reading intentionality or goal-directedness from the actions of humans but are instead simply reading behavioral cues and drawing associations to further behaviors? I think there is a fairly simple and convincing way to answer "no" to these questions, and to do so, I return to the work of Mitchell and his collaborators, as well as some considerations from feminism and critical theory. These latter fields, though seemingly far removed from the discussion of play thus far, have made substantial contributions to phenomenological considerations undergirding human-animal bonds. As we saw in chapter 2, Donna Haraway's work, for example, supports the idea that we ought not to be seeking out a perfect mirroring of human thought in nonhuman animals, but instead a *matching*, which is precisely what we find when we examine play between humans and dogs.

Matching, Mirroring, and Critical Anthropomorphism in Human-Dog Play

Interspecies play is equally compelling and confusing. On the one hand, it seems patently obvious that understanding is occurring between, say, dog and human, in a game of fakeout or avoid-fakeout. However, we should caution ourselves against assuming without solid reason that the dog conceptualizes intention or deception in the same way as the human. If a human continually "fakes out" a dog, and we see the dog's behavior change appropriately to avoid being duped, it is plausible that the dog has formed new associations and has learned from the interactions. An uncritically anthropomorphic stance would lead us to think the dog knows the human intends to deceive and is thereby avoiding being tricked. This reading assumes too much. The dog need not know that the person intends to deceive in order to avoid fakeout; instead, the dog can associate patterns of behavior and contingencies of the game in order to predict future behavior. In other words, just as we discussed in chapter 3, a dog is not required to have a robust theory of mind to meaningfully play with a human. Nevertheless, to claim that the dog understands nothing about the intention of the human in play would be to go too far in the opposite direction on the anthropomorphic scale and to perform what Andrews and Huss (2104) have called "anthropectomy," or removing any and all of the human-animal similitude from the account. That humans and dogs can and do play so successfully suggests that there is some kind of *interspecies intentionality* present and that dogs, just like humans, are aware of it. I turn now to the work of Mitchell and his colleagues, because I think their work on dog-human play provides a convincing picture of that space between uncritical ascription of human mindreading skills to dogs and unfair anthropectomy that overlooks important ways in which dogs do in fact understand human intent.

Mitchell and Thompson (1991) conducted a study in which they paired dogs with unfamiliar persons and observed the modes of play that emerged. The dogs were also accompanied by their guardians, who simply stood by and observed the play, with the idea being that this would provide enough familiarity to the dogs to comfort but not distract them. As with previous studies Mitchell and colleagues had conducted, in this study the researchers were interested in "social play" because, unlike with more structured games or other activities with clearly defined goals, the "goal" of social play is, in some sense, the play itself. Earlier I mentioned this aspect of play in terms of what Mitchell and Thompson (1986b) refer to as its *center-directedness*, or what Piaget (1945) refers to as the *autotelic* nature of play. Another way to think about goals in play is that they are repeatedly attained, with goal-directed actions serving both to reach those ends and to ensure that the goal is continually reinstated. Thus, for the purposes of this study, Mitchell and Thompson were specifically seeking to determine what sorts of goals were present among the players and how those goals shaped the interaction.

As we discussed earlier, play is arguably composed of projects and routines, even if those elements are not explicitly thought about by the players. A project, according to Mitchell and Thompson, is characterized by "repetitive action sequences during which a player calibrates its control over something," and routines are the interactions of projects of the players, either in their simultaneous occurrence or immediate succession (1991, 196–98). Thus a project might be something like "chase human/dog," "keep-away," or "fake-out," while a routine would involve a sequence such as "chase dog: chase object, chase person, object keep-away, run away, self-keep-away, surrender." The routine here described involves a series of compatible projects repeated and enacted at various points to maintain the overarching goal of "chasing the dog." Human-human play is similar in many respects. When I am chasing my toddler around the house, he will often turn around and charge at me, so as to chase me, thereby changing projects but maintaining the routine generally. I might then hide from him, jump out and scare him from my hiding place, and then begin my project of chasing him again, until one of us finally surrenders to the chase. In this sense, projects shift and change and routines are sustained by these transformations, with the overall goal being "to chase."

When the human-dog dyads in this study were set to play, they were given several objects such as balls, cloth, and rope and told to simply play in whatever manner and for however long they felt like, the intent being for the researchers to observe how humans and dogs play together. The idea was for the play to be spontaneous and for the researchers to record ways in which projects and routines emerged. Mitchell and Thompson found that routines were much more often the result of compatible projects than of incompatible ones. So, for example, the person-initiated project of "chasing the dog" combined with the compatible dog-initiated project of "run away" naturally lent itself to a sustainable routine, whereas if a person initiated "throw object" and the dog initiated "chase person," it was far less likely that a routine would result, and if it did, it would not last as long as in the case of compatible projects. The researchers also found that enticements were integral to the play projects, which included things like self-handicapping. For example, in a dog-initiated game of object keep-away, the dog might bring the object closer to the person, so as to almost allow the person to grasp it, thereby enticing the person to continue engaging in the game. There were also manipulations. In one case a woman who was trying to get a ball from a dog playing object keep-away pointed to another item and said, "Look, a stick!" and picked it up, enticing the dog to drop the ball and try to get the stick, at which point the woman snatched up the ball. Dogs were far less likely to respond to suggested projects than were humans, and humans were the only ones to use projects manipulatively, as in the case of "fakeout."

This last point is perhaps the most philosophically intriguing element in the study, as it points to the need to appreciate important differences among humans and dogs in terms of their experiences and perceptions within play interactions. However, the data collected in this study also point to the need for nuanced descriptions of behaviors, rather than assuming an all-or-nothing approach. Although Mitchell and Thompson found that dogs did not use manipulative strategies to deceive their human partners, they did engage in projects that have deceptive elements, such as keep-away and self-keep-away. Furthermore, dogs were inclined to engage the project of "avoid fakeout" when humans initiated it as part of a routine. As Chevalier-Skolnikoff (1986) has argued, it might be better to think of dogs as being aware of the contingencies embedded in each project or routine, rather than having a conception of deception itself. That is to say, dogs who hone their skills at avoiding being duped by a fake throw are not necessarily reading the mind of a human who is intent on deceiving, but they recognize that the supposed action of throwing is not going to come to fruition, and thus, in order to avoid being "faked out," they adjust their own behavior accordingly. As Bekoff and Byers (1998) argue, it is not the case, despite what some behaviorists might claim, that the play in which dogs are engaged is meaningless or "purposeless," even if dogs do not perceive deceptive actions with the same depth as their human counterparts.

Mitchell and Thompson see the results of their study, along with other studies that show similar results, as vindicating the need for what Ryle (1968) and later Geertz (1973) referred to as a *thick description* when characterizing animal behaviors. To understand a thick description, it is helpful to juxtapose it with its opposite, an *atomistic* description. One might worry, for instance, that it is already too much of an assumption to code data from studies such as the one just described in such human terms. Rather than describing a dog's actions as part of the "project" of "keep-away," a more atomistic description would instead code the behaviors as discrete movements, such as "picking up the ball," "moving with ball away from human subject," and "continuing to maintain a certain distance from human subject." This way, we are not guilty of uncritical anthropomorphism. But, as Mitchell and Thompson note, in the case of dog-human play, refusing to use these thick descriptors such as "chasing" or "avoiding fakeout" is tantamount to watching someone engage in the discrete actions of "cracking eggs into a bowl, whipping them together, pouring milk into them, stirring this mixture together, and so on" but insisting that "making an omelet" is an unwarranted thick description of these basic actions (1991, 214). Furthermore, it is rare to find anyone worrying over the use of thick descriptions in human behavior, because it is largely accepted that mentalistic terms pick out real phenomena in humans. However, as we have seen, there are reasons to be skeptical, at least in part, of some of our folk psychological terminology and what we are actually describing when we talk about human behaviors. Moreover, the very idea behind utilizing thick descriptions is that they involve not only the behaviors themselves but the contexts in which those behaviors occur. The action of running, described in isolation, is vague and could mean "getting exercise," "attempting to flee a fearful situation," or "playing a game of tag." If all we are ever permitted in ethology are atomistic interpretations, then all we can ever hope to explain are basic physical movements. Atomism would even resist calling interactions "play." So, in the case of dog-human interactions, running with a ball in one's mouth is the extent of our descriptive power, instead of the much more plausible and intuitive interpretation of the action as, say, part of the routine of "fetch." If we see that the dog is performing this action repetitively in conjunction with bringing the ball close to a human, dropping it, and then running toward it after a human throws it, in some sense what we decide to label the action is arbitrary. Call it what you will; there is something genuinely playful going on, and that play is not merely haphazard, unintentional, repetitive behavior with no purpose.

Dennett's work is again relevant here. In "Real Patterns" (1991) he addresses the worry of thick descriptions, albeit from a different angle, and he does not use the term "thick." Instead he is grappling with the question of realism as it pertains to folk psychology. He wonders whether the word "belief" picks out a real phenomenon, or if it is merely a label we use to explain observed behaviors. For instance, if I know that my dining companions are vegan, and I see that they refuse the bread as it is passed around the table, I might say that this is because they believe that the bread was made with animal products. And if I find out that indeed the bread was made with eggs, my explanation of their behavior seems even more warranted. But, as Dennett reminds us, there are philosophers—much like the behaviorists/atomists Mitchell and Thompson are opposing—who think that terms like "belief" are placeholders for basic actions, neurobiological processes, and past conditioning that are more accurately described as such. Eliminativists such as Churchland (1992) think that we ought to replace all this folk-psychological language with more objectively measurable language found in the sciences, so that instead of describing someone as "believing" this or that, we can talk about the person as behaving in such a way as to bring about a certain outcome, or rather than using mentalistic terms like "sad," "elated," or "angry," we should speak of serotonin levels, dopamine, and adrenaline.

Dennett, however, takes a pragmatic approach, and it's one that parallels what Mitchell and Thompson claim regarding the descriptions of projects and routines in human-dog play. Briefly, if the worry is whether we can prove once and for all that these thick descriptors or folk psychological terms pick out real patterns in the world, then the answer is probably "no." But notice that this skepticism applies to *all* attempts to characterize pretty much anything in "thick" terms. Am I really seeing a table in front of me, or am I just imposing that structure onto an array of atoms and particles arranged in a certain way? Do dogs actually understand that the person they are interacting with is not going to throw the ball this time, or have they simply been conditioned to respond to the stimulus in a way that reflects the past few failed throws? In other words, the ontological status of nearly all patterns we observe (or think we observe) and give names to can be questioned. If, however, we are not interested in abstract metaphysics, but instead in usefulness—how *pragmatic* are our choices in terminology—then it turns out that thick descriptions do a lot of work. Characterizing human action in terms of beliefs and desires allows us to predict with a significant amount of success what actions will result from those mental states. Likewise, as we have seen in the case of describing human-dog play interactions, utilizing thick descriptions allows for a much more nuanced understanding of the way two very different species of animal can engage in meaningful play, understanding one another's actions as goal-oriented, even if the conceptualizations of the projects and how they factor into those goals are distinct.

Much as I argued with regard to mindreading capacities—that it is better to conceive of them on a continuum rather than as an all-or-nothing skill set—the discussion so far overwhelmingly suggests that in play we see the same continuum. As the study conducted by Mitchell and Thompson (1991) indicates, humans are much more aware of deceptive strategies, and they employ them in play routines, but this does not mean that dogs lack any understanding whatsoever of deception. The difference, it would seem, is in degree, not kind; while humans can form elaborate plans and simulations so as to enact a deceptive manipulation, dogs can match this with resisting being tricked. Rather than mirroring humans in their abilities, in play, dogs *match* humans quite well, engaging in joint projects, collective routines, and shared goals. As Mitchell describes it in a more recent paper on creativity in play, even though they are not "equal" in the sense that humans are more sophisticated in their strategies than dogs, "in the context of their social play, they appear to view each other as equals" (2015, 37).

This notion of partnering without perfect symmetry, or matching without mirroring, finds a parallel line of thinking in critical theory and posthumanist philosophy regarding the human-animal divide. I now focus briefly on the work of Donna Haraway and Vinciane Despret, though there are indeed many other important figures upon whom I could draw. My intention here is not to provide an exhaustive account of an entirely different field, but rather to highlight some of the ways in which disparate modes of inquiry can converge on similar ideas, and how, when they are brought into discourse with one another, even more clarity, and in some cases more confusion, surrounding a given issue can emerge. Most important, I want to examine the ways in which these thinkers problematize some of the conceptual categories deployed in thinking about and thinking-with animals, and how assumptions concerning objectivity might be hindering human-animal research.

In the study discussed above, Mitchell and Thompson (1991) draw an analogy between the type of communicative acts present in human-dog play with Mead's notion of a "conversation of gestures." Mead himself was interested in dogs, and used an example of dogs engaging in a fight to illustrate what he meant by a conversation of gestures: "The very fact that the dog is ready to attack another becomes a stimulus to the other dog to change his own position or his own attitude. He has no sooner done this than the change of attitude in the second dog in turn causes the first dog to change his attitude. We have here a conversation of gestures. They are not, however, gestures in the sense that they are significant" (1934, 42).

What Mead is saying here is that dogs, and many other animals, undoubtedly communicate, even if they do not know that they are communicating. In other words, their gestures might not be intentionally put forth in the same why I carefully plan and construct a string of linguistic utterances I might be conveying to my students. In such a case, I am aware that I am communicating and aware of what each of the symbols or gestures ought to convey. For dogs, however, as Mead argues, this metacognitive level is not present.

Perhaps an even better way to think about all of this is to consider how the dogs are *dynamically signaling* to humans and that this is a genuine form of communication. The signals dogs give—a play bow to entreat the human into a game, for example—need not have intention behind them. As Maynard-Smith and Harper (2004) argue, much of animal communication may be nonintentional, but there are nuanced differences between the ways animals signal, and why some signals are more successful than others. Indeed, what the dogs are doing is not the same thing as what an orchid does when it emits a fraudulent scent to attract a bee to pollinate. The signals the orchid sends to the bee are nonintentional, to be sure, but they are also inflexible. They do not occur "on the fly" in response to the ever-changing behaviors of the bee, nor are they dynamically interactive, such that subtle cues from the bee or the flower can drastically change the course of the exchange. The orchid-bee relationship is relatively fixed, and so is the signaling. In dog-human play, however, the signaling is much more flexible. The idea of dynamic signaling fits well with the continuum of social cognition abilities discussed in chapter 3, as in the case of human-dog play we can see definite signs of fundamental social cognition exhibited by the ever-evolving signals given by dogs as they respond to human cues. Likewise, signaling is generally accepted nomenclature for ethologists who might be wary of adopting terminology like "conversation of gestures" to explain human-dog communication. In my estimation, it can be termed dynamic signaling or conversation of gestures, and that matters less than the acknowledgment that what is taking place has meaning and significance for both species, as it is part of a game the two are playing.

In her 2008 book *When Species Meet*, Haraway tackles the issue of companion species and conversations between them and "us." She examines interspecies communication from a host of angles, but the one I find most compelling and germane to this particular discussion is her elucidation of how humans and dogs play the game of agility together. In the sport of agility, unlike the free play Mitchell and Thompson asked participants to engage in during their study, there are prespecified rules, at least insofar as the human participants have an idea of what they want the dogs to do—jump through a hoop, climb an A-frame, and other tasks. In competitions, the rules are even more important. When climbing the A-frame, for instance, the dog's feet must touch a certain spot, usually indicated by a differing color, near the bottom of the structure, on both sides. If these rules are not followed, the pair will lose points. If the agility is just "for fun," as it is at my house—we have some equipment in the backyard, and my dogs and I go out there just to run around and get exercise—it is still the case that the play is more structured than if I casually pick up a rope and begin playing keep-away with my dog Darwin, enticing him to then engage in tug-of-war with me. The projects in agility are predetermined—"jump over gate," "sit on the pause box," and "run through the chute." But, as Haraway correctly notes, when involved in the play of agility itself, communicating to the dog that you want to engage in this or that project is usually not a linguistic act. Or it is, at best, minimally linguistic. A simple nod or flick of the wrist can indicate "run through that tunnel," and though I might say, "Hoop!" to indicate I want my dog to jump through a hoop, it is far less common to hear words on agility courses. The conversation, it turns out, is in the gestures.

So far, I've described communicative acts only from the perspective of the humans, who arguably understand that their gestures have significance—although, in the case of so much of dog-human play, even in agility, from my anecdotal perspective, how the gestures come to signify what they do is a process that emerges through the interaction, much in the way I argued in chapters 2 and 3 that thinking, according to the "radical" views in philosophy of cognitive science, just is interacting with others. What about the dog's perspective? Haraway points out that it at least seems as if her dog enjoys the game for the game's sake, and most agility enthusiasts will probably say the same of their dog-partners. But why? Perhaps it is because of the subtle cues dogs give to indicate satisfaction, such as when my dog Darwin has run through our backyard course "successfully"—which means he has enacted projects in the way I wanted him to—and he immediately runs to me, jumps up and puts his paws on my stomach, and cocks his head to one side as if to say: "I did good, huh?" Obviously, I am guilty of far too much anthropomorphism in my interpretation of Darwin's actions as bearing linguistic significance like that, but all the same, it is not a far cry to watch dogs competing in agility games, or simply playing in nonstructured ways, and see the happiness and pleasure they exude through their actions and gestures. These conversational gestures, however, make sense only insofar as they are part of the larger whole we can think of as the "game." As I have argued throughout this book regarding how we ought to understand cognition generally, the meanings, emotions, and gestures within human-dog play cannot be understood without reference to the context in which they are occurring.

Nor, as Mitchell and Thompson argue, can we expect that an atomistic interpretation of behaviors will do us any favors when trying to comprehend meaning in play. Haraway's account of communication during agility emphasizes this point, though it is worth reiterating that it's not entirely clear what dogs are intending—if they are intending at all—to communicate with us in agility or any other play. And yet, examining agility play uncovers something peculiar about intentionality and mindreading: dogs, or at least *some* dogs, know very well what we want and what we intend, as is evidenced by their obedience during and successful completions of agility courses. The philosopher Despret, in examining the relationships between breeders of domesticated farm animals and the animals themselves, found that the same communicative inequity existed. Breeders themselves were clear that the animals know very well "what we want, but we, we don't know what they want" (2008, 31). She found that unlike philosophers, breeders were uninterested in discussing how their animals were similar to or different from humans, but were rather interested in learning how different situations give rise to different needs, both for the animals and for the humans. I think this finding is crucial to the present discussion, to the study of human-animal relationships generally, and moreover in further problematizing some assumptions made in philosophy of cognitive science against which I've been pushing thus far.

In her 2008 paper "The Becomings of Subjectivity in Animal Worlds," Despret criticizes Dennett, whose arguments I've been mostly sympathetic to thus far. I think her concerns are not only warranted but are closely related to what she finds regarding communication between breeders and their animals. Dennett (1996) proposes a reframing of something Wittgenstein argued in his *Philosophical Investigations*, namely, that if a lion were to suddenly begin speaking in some human language to us, we would not be able to understand it. Instead, Dennett suggests, we ought to be wondering whether, if a lion could speak, would it be able to tell us anything interesting about lion-ness? Dennett's a priori answer to this hypothetical question is that if a lion were to have the capacity to speak qua human, it would no longer be "lion enough" to be a proper representative of lions. This argument in some ways parallels one we discussed in chapter 1 regarding Nagel's claim that to fully know what it's like to be a bat, or any other creature, we would have to actually be that creature. But of course, if I were to transform into a bat, I would cease being a human and would not therefore be a human understanding batness but would now just be a bat understanding batness. Dennett's proposal, and to some extent Nagel's, according to Despret, glosses over the complexities of what might count as a representative voice for the whole of a species and furthermore assumes that humans are fit to determine if this or that member of a given species is a proper exemplar. Dennett, says Despret, asserts that "we must ourselves be the judges of what it is that ensures that a lion has something to teach us on the subject of what it is to be a lion. That is to say, we have slid surreptitiously from the question of *representing* to the question of the *representative*" (2008, 127). This double "we," as Despret calls it—the "we" who analyses the "other" and the "we" who is so different from other animals—is, as I see it, at the crux of the problems persistent in ethological philosophy when it comes to navigating between a rigid anthropocentrism/anthropectomy and an uncritical anthropomorphism. For at once it assumes a uniformity to "our" experiences qua human and a uniformity to other species' experiences. To be fair, Dennett and, most assuredly, Nagel are well aware that subjective consciousness is unique to an individual, and it would be making a straw man of their arguments to assert that they think "the human experience" is in fact universal. Nevertheless, in treating other species as though one member might be able to "speak" for all its kind is problematic, to say the least.

Despret urges us to pay more attention to those human-animal dyads in which specific *apparatuses* are developed and sustained. By "apparatus" she means a scaffolding or framing of an encounter. As we have been discussing with regard to *play*, a project or routine would be an apparatus that structures the interaction between me and my dog and, more important, the types of communicative gestures we can use to convey meanings. Breeding is another apparatus, Despret argues, whereby the humans and animals are conjoined in a process of bringing forth and ending life, and hence, as she notes, the breeders are not interested in how the animals are essentially different from their own. After all, breeders and their animals literally live and die together, so such questions are not at the forefront. Instead, as noted above, breeders tend to want to know more about what specifically, in terms of breeding, their animals want, need, and intend. Apparatuses, then, not only serve to structure the interactions between humans and animals but also reveal *what sorts of questions even make sense to ask.* Here I quote Despret at length as she discusses parrot "language" and the notion of apparatuses:

> If the parrot can talk, we do not know what it is, nor what parrotness is, nor anything about the point of view of parrots on the world. But we do learn in a viable manner about its point of view on the apparatus. We learn something about its point of view on the new materials with which it will make a world: colour boxes, numbers, words, a grammar, forms, humans and abstractions. In the same manner that the refusal to talk, in other apparatuses, constitutes an expression of the parrot's opinion in relation to the relevance of what it is asked, the fact that it engages with, accepts and actively transforms what becomes a part of its world, translates an extension of this world and therefore an extension of its subjectivity as "parrot-with-human." (2008, 128)

This quote neatly sums up what I have been arguing regarding the context-sensitive nature of meaning and communication in play, and I think these are ideas to which Mitchell and Thompson would be sympathetic, given their findings. Moreover, the idea that subjectivity can be explained as a being-with, so long as it is scaffolded with the right sorts of apparatuses, is essentially the overarching argument of this book. I have been mainly interested to explicate how we *think-with*, rather than to attempt an account of subjectivity itself, but the gist of Despret's work is similar insofar as cognizing, meaning, and subjectivity—all of these things typically taken to be internal and subjective phenomena that are prior to and the conditions for communication—turn out instead be the result of those communicative interactions themselves. Play, as we have seen in this chapter, is perhaps one of the best examples of how situated and contextualized conversations emerge between humans and dogs.

Conclusion: Outplaying Enactivism?

In her most recent book, *Staying with the Trouble: Making Kin in the Chthulucene* (2016), Haraway draws some similar conclusions to what I am arguing regarding our relationships with other animals, although her purpose in the book is perhaps more noble than my own here. In characteristic Haraway fashion, she dismantles dualisms and confuses categories so as highlight the messy and complex relationship humans have with all the other species inhabiting "Terra," or planet Earth. She is hoping to find a path to better living-with those other species and making a world that is habitable, rather than wallowing in what she sees as the false dichotomy of hope versus despair. We must, she argues, "stay with the trouble" and reorient our thinking. My project so far has been to try to get philosophers and ethologists to think about thinking through the lens of human-dog dyads so that we might gain a more diverse perspective on cognition, though I do like to think that my project is in some small sense a part of the more global project Haraway has envisioned. Admittedly, I have always found that I agreed with nearly everything Haraway has argued over the years, but some of these recent arguments have been more difficult to accept, especially her "attack" on autopoiesis. Autopoiesis, recall, refers to the self-organizing and self-sustaining nature of organisms and is a central concept in much of the enactivist arguments in philosophy of cognitive science. Autopoietic creatures, while homeostatic and predictable to some extent, are also constantly interfacing with their environments and re-creating themselves by doing so. Thinking, according to some versions of enactivism, just is this process of self-preservation mixed with self-augmentation within one's environment. Thus, to read Haraway claiming that "Bounded (or neoliberal) individualism amended by autopoiesis is not good enough figurally or scientifically; it misleads us down deadly paths" (2016, 33) certainly gave me pause to reflect and examine my enactivist commitments.

Instead, Haraway argues, we ought to be thinking in terms of *sympoiesis*, or a *making-with*. The term was first used in a master's thesis in environmental studies by M. Beth Dempster in 1998. Haraway notes that in it Dempster argues that many systems that appear to be autopoietic are in fact sympoietic. That is, they are "collectively-producing systems that do not have self-defined spatial or temporal boundaries. Information and control are distributed among components. The systems are evolutionary and have the potential for surprising change" (quoted in Haraway 2016, 33). I think a sympoietic system characterizes quite well many of the human-animal encounters we have discussed so far, and particularly the human-dog dyad in play. "System" used in the singular will likely confuse some readers, because the knee-jerk reaction is to proclaim that a human-dog dyad is ultimately *two* systems—indeed, two *autopoietic* systems if you are a committed enactivist. But here Despret's work is incredibly helpful again. Certain interactions are such that *subjectivity itself* emerges from the exchange. In play, for example, I am not a human playing with a dog, but rather I *become* human-with-dog. Sympoiesis, in the way Haraway uses it throughout the book, further means a cocreating. Neither I nor my dog "makes" the game—we collectively create it as we go. And as we have seen so far, the communicative acts that ensue within that apparatus/project/context are meaningful precisely because they are part of the game.

This brief discussion of sympoiesis is meant to entice the reader to one final chapter in which I attempt to tie these threads all together and revisit the overarching argument of the book. We have discussed play extensively in this chapter, but have only begun to scratch the surface of how play is "creative." Creativity, in play, in sports, and in thinking, is the fulcrum on which my argument for an enactivist approach to human-canine coactive cognition hinges. Thus, part of the next chapter will address how we "make-with" dogs, in play and in other interactions.

Of course, if I am to accept Haraway's argument for sympoiesis instead of autopoiesis, I will need to attempt to reconcile this with the predominant enactivist framework, as there is certain to be pushback against the idea that at least sometimes, dogs and humans are not autonomous units that are spatially and temporally circumscribed. As I will discuss in the final chapter, a reason for retaining at least some kind of bounded and encapsulated notion of organisms—even if they are dynamically open and always interacting with environments and other species—is that we can *predict* a great deal about their behaviors. To put it differently, if there are not singular organisms that interact with environments, it is hard to meaningfully predict that this or that environmental context will yield this or that behavior, as the boundaries have been dissolved between the two to begin with. I think, however, that keeping Despret's arguments regarding apparatuses in mind as we proceed, along with some careful considerations of predictive capacities, how they are afforded, and what they afford, will be key to integrating sympoiesis into my overall enactivist account of human-dog coactive thinking and creating.

CHAPTER 5

Dynamic Duos

MAKING-WITH, THINKING-WITH, AND
ENACTING INTERSPECIES COLLABORATIONS

Creative Dynamics of Thought and Action

So far, we have seen ways in which cognition can co-emerge during all sorts of interactions: between two or more humans, between humans and tools, and between humans and nonhuman animals. In many parts of the world, dogs and humans have closely bonded over the last fifteen thousand years, and this bond has afforded them special ways of cognizing. As I argued in the previous chapter, these unique ways of thinking are exemplified when the two species *play together*. We further saw that "play" is best conceived as an "apparatus" or scaffolding by which interactions generate meaningful communication and cognition. These meaning-generating interactions are sympoietic, insofar as they involve two or more beings tightly coupled that form a unit from which the thoughtful action arises. This account dovetails nicely with the enactivist framework I have been defending, and we can even find useful analogues in biology—such as the *holobiont*—to further bolster this idea of sympoietic interaction.

Some concerns have arisen, however. For one, part of the allure of an autopoietic system for enactivists is that it is stable enough to make some reasonable predictions about the system's behavior. On the other hand, a sympoietic system, by definition, comprises ever-shifting conglomerations of multiple organisms in world-making, so how can we reasonably expect any predictable regularity? Perhaps I must abandon all claims that I am indeed defending an enactivist account of cognition, if I am so strongly committed to sympoiesis as the framework explaining interspecies dynamics.

Sympoiesis and enactivism are not irreconcilable, in my view. In fact, it is my contention that many enactivists, when discussing *social dynamics*, are just describing sympoietic systems *at play*, even if they utilize more autopoietic terminology. In discussing play, we haven't considered a fundamental element present in most if not all forms, whether spontaneous or more rule-based, and that is creativity. Creativity is closely tied to the idea of *making* in sympoiesis, as it often involves collaborative effort. We often think of creativity as an individual phenomenon, such as in the case of a brilliant artist painting an exquisitely beautiful picture. To be sure, this is one mode of creativity. This version is not at odds with a sympoietic understanding of creativity, either, as I explain in this chapter. To do so, I turn back to the question of how to reconcile the predictability and stability of autopoietic systems with the seemingly unpredictable and unstable nature of sympoietic systems. I further examine how creativity functions in interspecies dyads and how this can potentially shed light on reconciling the supposed tension. Then I examine some recent discussions regarding prediction itself—in particular Andy Clark's 2016 argument for a *predictive processing* (PP) view of cognition. I argue that although a dyad such as human-with-dog might at first appear too unstable to count as a meaningful unit for predictive inferences, if we closely examine the innovative aspects of the interaction, we can see how these regularities reliably emerge as the interaction unfolds.

To bolster the argument that enactivism is not at odds with my sympoietic account of human-dog interactions, I follow closely the arguments that Shaun Gallagher and Micah Allen put forth in a 2016 paper where they attempt to reconcile PP with enactivism by instead arguing for *predictive engagement* (PE) as a model of cognition. And finally I return to interspecies collaborations, now armed with support from these enactive-predictive models, to further defend the idea that such views actually *entail* a sympoietic account of human-dog coactive cognition.

Innovative Play, Creative Cognition, and Making-with

The spontaneously generated play that emerges between humans and dogs, much like the play that emerges among small children, is both *center-focused* and *goal-oriented*. On the one hand, play is just "for itself"; the point of playing is, as Burghardt describes it, *autotelic*. The for-its-own-sake-ness of play, however, does not rule out a purpose or end to play. As we discussed in chapter 4, it is well established that play facilitates important developmental trajectories for human children, even if they are unaware that such learning is taking place. The same is true of play between humans and dogs, though the specific ends are most likely different. The spontaneous play Mitchell and Thompson (1991) observed allowed for new projects to become routinized and for those routines to give way to increased understanding of intent to act, thereby allowing for deceptive strategies to emerge, at least on the part of the humans. For the dogs, though they might not engage in intentional deception the way humans do, they certainly learned from the interactions how to avoid *being deceived*. In keeping with the idea of a continuum for capacities that we might generally place under the heading "social cognition," in play, we can see dogs most certainly sustain *some* type of rudimentary mindreading capacities. The honing of these social-cognitive skills then further serves to sustain the play. We might thus plausibly speculate that some of the "goals" of play—even in spontaneous and mostly unstructured free-form play between dogs and humans—are to become acquainted with the play partners, to learn something about them, and to become more attuned to their thoughtful actions. What we have yet to discuss is how play, whether it is unstructured or rule-based, almost always has a creative or innovative element to it. This aspect, I will argue, further demonstrates why we ought to take play more seriously if we hope to uncover more about human-dog interactions specifically, as well as nonhuman animal cognition generally.

Mitchell characterizes participants in human-dog play as "part of a collaborative dyad in which each player tries to gain and retain expertise in his projects within the accepted constraints of the game" (2015, 33). Part of what makes it possible to "gain expertise" is to learn new strategies and find novel and more efficient solutions to problems within that game. Thus, when I speak about creativity or innovation in this section, I mean the very general sense in which one adapts to the dynamically shifting features of the play space, including the "possible moves of the adversary." Of course, creativity might come in the form of seemingly blind insight, the likes of which we tend to attribute to artists or scholars who have almost accidentally stumbled upon the next most beautiful sculpture or eloquent argument. I am less interested in that type of creativity, though I have a hunch that it can be explained via the same guiding principles with which I characterize creativity in play, albeit with perhaps more complexity and nuance. Nevertheless, I want to avoid a digression into the ontological status of what makes something a creative act. There are countless arguments pertaining to this issue, which goes far beyond the scope of this book. For simplicity's sake, therefore, when I refer to the creative or innovative aspect of actions within play or any other activity, I am referencing the capacity to solve problems efficiently or to better position oneself in the game by trying out new projects or changing the routine ever so slightly. In short, to be creative is to be in the mode of *making, innovating, or improvising*. It could be as simple as a human utilizing a fakeout technique to trick a dog into chasing an imaginary disc, until the dog catches on and instead jumps to snatch the disc from the person's hand as it is being surreptitiously hidden behind the person's back. This is a creative solution within the confines of the "problem space," which in this case is what we might just call a "game of fetch."

The definition of creativity I am utilizing captures how this element is present in other forms of play among humans, such as in sports. In fact, Mitchell (2015) has already done the work of connecting creativity in human-dog play and human sports. For instance, he characterizes play that he observed between Chris and Hercules, a human and a dog respectively, as fitting the parameters of the most common definition of creativity in sport, namely, the ability to create work that is, as described by Sternberg and Lubart (1999), both novel—that is, original and/or unexpected—and appropriate, meaning useful. For example, in the fakeout/avoid-fakeout routine that emerged between Chris and Hercules, Mitchell notes that the norm of "escalating reciprocity" was operative: the more Chris varied and tested new fakeout strategies, the more Hercules became attuned to these maneuvers, and as he "improved" his responses to Chris, Chris continued to attempt to better his reactions. While the two were creating novel moves that were unexpected, these moves were still within the confines of the game.

This capacity to simultaneously surprise and operate within given parameters highlights another parallel that can be drawn between human-dog play and sport if we consider what *constraint theory* (see Elster 1984, 2000; Lewandowski 2007) says about creativity. The idea is quite simple: the constraints placed on players in a game are *part of* the creative process itself. To maximize one's capacity in a sport like swimming, for instance, one must use the constraints of the game as guiding principles, and even more so as the determining factors for what would even count as a "good move" in the first place. There are rules about what strokes are permitted, how many hands must touch the wall upon turning around, and so forth, but excellent swimmers find subtle ways to capitalize on the push-off, to minimize drag, and to use the currents they themselves have created to propel their bodies through the water faster.[1]

The example of swimming I chose to illustrate constraint theory is much more about the way a player might utilize parts of the physical confines, in this case the pool, for "constrained maximization." Of course, for our present discussion, sports where "collaborative dyads" are constitutive of the game are much more compelling, but I do think it's worth pointing out that the relationship between constraint and creativity can be emphasized even in cases of organism-nonorganism dyads. I return to this idea later in the chapter when we broach the issue of predictive processing. When it comes to intersubjective dyads, on the other hand, as Mitchell notes, the environment now includes another person or dog who has intentions to act and goals within the parameters of that game. Thus "the other's actions are not only required, but offer the possibility of creative responses to these actions" (2015, 35), which is precisely what he observed in the exchanges between Chris and Hercules. With regard to human sports, boxing provides a nice illustration of constraint theory, as both Elster (2000) and Lewandowski (2007) discuss in their respective studies. Not only do the boxers provide actions that allow for creative responses within the boxing match, but they do so with the shared goal of improving and maximizing their abilities within that space. Lewandowski suggests that boxers "engage in a form of shared cooperative action and practical improvisation designed to instruct one another in mutually beneficial ways, such as when boxers reflexively correct one another's mistakes with controlled well-placed boxes" (35). This idea of "practical improvisation" is one I want to explore a bit further, as it neatly ties together many important concepts that are central to the overarching argument I am defending.

Improvisation is closely related to creativity, as it is often in improvising that novel solutions, moves, or plays are created. We might say that improvisation is itself a form of play—a *playing-with* boundaries while also remaining within certain parameters. Jazz musicians and comedians are two notorious types of artists who "improv" or perform "on the fly." However, improvisation need not be relegated to the arts. As we discussed in chapter 3, coordinating movement and co-attuning affect, such as those processes that occur between parent and child, are improvisational—that is, they are not the result of choreographed or preplanned action. Instead, they arise through the dynamic exchanges taking place in the here and now, and yet they are also part of systematic development, so have a fairly predictable general trajectory. Conversations are often improvisational in this way as well. So too do we find improvisational techniques in human-dog dyads—this is exactly what Mitchell (2015) describes when he records the evolution of the fakeout/avoid fakeout strategies deployed by Chris and Hercules. I want to take it a step further and suggest that improvising is a hallmark of cognition generally, and when it occurs in an intersubjective or interspecies context, it exemplifies what Haraway refers to as *sympoiesis* or making-with.

One other place we can see intersubjective improvisation and thought seamlessly intermingling is in *dance*. To be sure, some dances are far more choreographed and have more rigidly prescribed movements than other forms. Improvisational dance, therefore, is going to look much more like the other examples of sympoiesis described above. I argued something close to the idea that we can find sympoietic or "improvisational thinking" in a 2015 paper, "Thinking-Is-Moving: Dance, Agency, and a Radically Enactive Mind." There I focused not simply on improvisational dance but on *contact improvisation*, and I argued that the creative elements in it are akin to what goes on in *participatory sense-making*. Recall, from chapters 2 and 3, that participatory sense-making is defined by De Jaegher and Di Paolo as "the coordination of intentional activity in interaction, whereby individual sense-making processes are affected and new domains of social sense-making can be generated that were not available to each individual on her own" (2007, 497). In contact improvisation, as the name suggests, dancers do not have choreography to follow, nor are they expected to move in any particular way.[2] The only "constraint" on the activity is that they must stay in physical contact with their partners the whole time. In improvisational dance generally, as in any other type of improvisational art or movement, constraint theory factors into the creative dynamics: though the dancers can choose to do whatever they want whenever they want, the movements must still show up as *dance* movements, and there are limits to what sorts of moves fit this constraint. In contact improvisation, however, the further constraint of maintaining partner contact is in place. This "rule" is not merely a limiting factor. If we recall how Mitchell speaks of constraint theory in terms of "escalating reciprocity," the contact required between the dancers is also a *vehicle* of creativity. The contact between the dancers is also an example of what Clark (1998) calls "continuous reciprocal causation," as it is what drives along the process of the collaborative dance. I quote Mitchell here again, when he discusses how others serve to both constrain and help create in the interaction: "the other's actions are not only required, but offer the possibility of creative responses to these actions" (2015, 35). Indeed, it is the contact between the two dancers that is at once the constraining and creating factor for the movements. I sense what my partner is doing through haptic awareness, and the reverberations of those movements shape and transform what is possible for me to do. An arm draped over my shoulders limits my capacity to jump, but it might afford me the movement of melting into the floor, thus enacting a series of conjoined moves on the floor and back up to standing. And, as I argued in my 2015 paper, despite the extemporaneous nature of contact improvisation, the movements are not mere "thrashings about"—they are, put simply, "dance moves." They have meaning. The meaning, however, emerges and is sustained in the contact, in the biomechanical feedback, and in the bodily communication afforded by this exchange.

If two dancers can cocreate in contact improvisation or *make-with*, so too can dogs and humans form sympoietic systems, whereby the activity they are engaged in is created and sustained in the interactions while also being constrained by the nature of the game or the rules of play. And as Despret argued regarding subjectivity in these cases of co-making, it is better to think of the participants in a conjoined or intersubjective union, such that when I dance with my partner, it is Michele-with-partner, or when Chris plays fake-out with Hercules, it is Chris-with-Hercules who is playing the game. The unit that has meaning for the sake of understanding the game itself and its constraints, in other words, *just is the collaborative dyad*.

The idea that it is not easy to decouple players during a moment of intersubjective improvisational play shows up on a much smaller scale as well. As I have noted, the *holobiont* concept, usually referring to a system comprising host + microbiota, bolsters this argument. What might seem like abstract and impractical metaphysics turns out to be concrete reality for many who study such complex systems. It is well established that humans are host to a vast array of bacteria and viruses, some of which are not only beneficial but necessary to our health and survival. The types of gut bacteria that a human hosts, for example, will change over time, but there is always some combination of human + bacteria, dynamically interacting. In fact, to be a human being, in many ways, *just is* to be a complex system of bodies tightly coupled. Thus it does not really make sense to say a human + bacteria, as the human already is the holobiont containing the bacteria. In other words, what it means to be an individual is not so simple, and this is not just a contrived philosophical thought experiment. Similar confusion arises when we consider "superorganisms." Leaf-cutter ants, for example, spend their time seeking out specific leaves that happen to produce a fungus when chewed. A range of fungi comprise the various substrata, but they all belong commonly to the family *Lepiotaceae*. The ants set up "gardens" to cultivate the fungi, which they can then feed on, and they meticulously care for the health of the fungi by keeping the growing areas clear of decayed matter, adding leaves, and deterring pests. These ants are also holobionts in their own right—they have a bacterium that grows on their bodies that produces antimicrobials that, when secreted, protect the fungi. There are even complex interactions among the ants themselves, as some "hitchhikers," who ride atop other ants while foraging, serve to fight off would-be destructive parasites (McVean 2019). This symbiotic "supersystem" constitutes a superorganism, according to Shik et al. (2018) and other researchers. At the very least, it belies the notion that individual organisms can be easily decoupled from their ecological niches.

As this brief foray into the close symbiosis among ants, fungi, and bacteria illustrates, disentangling the various "bionts" from the larger system they comprise—the holobiont—is not only difficult, but to do so would hinder us from a comprehensive ethological understanding of the ant's umwelt. If we are to even broach the subject of whether ants can, for example, reason about the best strategies for fungal gardening, it behooves us to recognize the very complex nature of what it means to be an ant who is inextricably tied to a variety of other life-forms. When we think of humans and dogs in this way, of course, the analogy to holobionts will only go so far. I am not coupled to my dog the way I am to my gut bacteria, to be sure. However, it is quite compelling to find out, for instance, that the intestinal bacteria my dog is host to very closely resemble my own internal flora. Human gut biomes resemble dogs' far more than they do pigs' or rats', which are traditionally studied to better understand our own bacterial flora. Scientists such as Coelho et al. (2018) are now directing more attention to a biome "right under our noses"; once again, the idea that we share more than a house with dogs is taking hold in surprising contexts. In the context of the argument I am making about sympoiesis, we might say that both dogs and humans are their own complex sympoietic systems, which are often engaged in tightly coordinated sympoietic interaction with one another. Although not exactly a "superorganism" like the leaf-cutter ant + its bionts, human + dog systems often share some striking similarities.

Sympoiesis arguably occurs on a much larger scale as well, such as in sociocultural exchanges. Although, to my knowledge, no philosopher of cognitive science has ever applied this concept per se, we nevertheless see places where it could be effectively utilized. Gallagher (2013) has argued that certain social institutions come to constitute forms of cognizing, such that when we are thinking about, say, a legal problem, and are utilizing the entire legal system to work through the issue, we are not really thinking entirely *in our heads* or all alone. Instead, our thoughts are subtended by what he calls "mental institutions." These institutions allow for improvisational dynamics, but they also operate within sociocultural constraint. If we think of language as a similar sort of social institution, then, as Clark has recently argued, these cultural media are not just outputs of our computer-brains, but are in turn shaping how we think and what we can *make-with* those symbols. As he puts it, words can "modify both what top-down information is brought to bear, and how much influence it has at every level of processing" (2016, 283). Finally, Heyes ties all these ideas back to creativity by reminding us that the capacity to innovate is itself a *product* of cultural innovation. Think of the way technological inventions are developed. There are sociohistorical constraints on what can be discovered or invented—email would not have been possible without computer "culture"—but with each innovation comes greater capacity to further innovate. Cultural innovations, in other words, "are products as well as producers of cultural evolution" (A. Clark 2016, citing Heyes 2012, 2182). And as Clark points out, these innovations are acquired not solely from the biological adaptations enacted in organism-world transactions, but via social interactions as well.

We have strayed far from dog-human play and creativity, but the thread is not lost. There is a worry we must address, however, as the discussion has shifted now to what we might think of as distinctively human capacities to *make-with*—performing arts, language, and so forth. Recall, though, my argument has been that cognition is not just to be found in these sorts of exchanges, but rather exists on a continuum, and is marked by organisms adaptively strategizing in their world-engagements and transactions. A question we might have then is *Where do human-dog sympoietic interactions fit on this continuum?* We might even wonder if human-dog dynamics—or any other interspecies exchange, for that matter—should count as properly cognitive. Perhaps what we are explaining when considering human-dog play is indeed closer to the ant-fungus scenario than I might wish it to be. In other words, we are faced with an objection to enactivism generally, namely, that it is guilty of defining cognition in such a way as to render it meaningless. It is just what occurs when organisms interact with their environments or other organisms. I will respond to these concerns in the next section, as I think they can be assuaged by closer examination of creativity in play and by a consideration of how not all cases of organism + world engagement will qualify as cognitive.

Another worry is that we ought to take care with how liberally we apply metaphors, as in the case of using sympoiesis to explain every form of dyadic exchange, or in characterizing all collaborations as "dances." I bring this last point up specifically because it pertains to the concerns raised regarding anthropomorphism. Ken Cheng notes that ant behaviors have been referred to as dance-like on numerous occasions, with authors using "pirouettes" to describe "saccadic turning movements with stopping points, and voltes, which are tight turns without stopping points, in the learning walks of three *Cataglyphis* species" (2018, 7). Though it might be apt to compare the ants' movements to dance, Cheng warns against comparing their behavior too closely to *improvisational* dance; their "scanning is a stereotyped behavior that contrasts with the variable and creative nature of improvised dancing" (ibid.) Furthermore, the saccading movements of the ants arguably serves a survival function and is fairly predictable—ants engage in the behavior more when in unfamiliar territory and alter their course depending on what the saccading affords their visual systems. The "ant dances," in other words, are well choreographed, while in contact improvisation the whole point is to eschew any need for choreography while still engaging in a creative art form.

This supposed dichotomy between predictability and creativity is one that I want to challenge in the next section. There are some compelling arguments that suggest the (human) mind is a predictive processing engine, and hence a lot of what might turn out to mark cognition is the ability to make accurate predictions about the world and to act accordingly. This view of the mind has been pitted against the enactivist framework for several reasons. For one, it goes against the framing of intelligent systems as *autopoietic* or self-making/creative. As well, it might push too much of the mind "back in the head" and back into the paradigm of standard, computational cognitive science. I address these tensions, as well as the ones I raised above, as we proceed. First, however, I want to examine predictive strategies and what they really imply about cognition. Cheng is correct that we should resist analogizing ant and human movement too much in terms of creativity—indeed, ant "dances" are often far more predictable than human contact improvisational dance—but this difference is in *degree, not kind*. As it turns out, predictable patterns or "routines" can and do emerge from dynamic exchanges between and among organisms and their environments, including those between dogs and humans, as well as ants and fungi. Those transactions often result in genuinely cognitive or intelligent processes that cannot be understood without all of their putative parts, even if those "parts" turn out to be multiple organismic beings. But this thinking-with and making-with is also flexible, innovative, and creative. Cognition, in other words, is predictably creative. As Cheng puts it, "Intelligence comes out of movements of partners on the fly, in loose cahoots with one another" (8).

Predictive Processing and Sympoietic Systems

Despite being well known for his "radical" views in philosophy of cognitive science, Andy Clark appears lately to have retracted a bit of his "supersized mind" by offering an account of cognition that is much more brain-based. In *Surfing Uncertainty: Prediction, Action, and the Embodied Mind*, Clark defends the *predictive processing* (PP) model of the mind, which is the idea, quite simply, that "brains like ours ... are predictive engines, constantly trying to guess at the structure and shape of the incoming sensory array" (2016, 3). The PP framework Clark endorses draws heavily on principles of Bayesian inference, which have a long history in the sciences of mind, as this type of inferentiality goes far to explicate key features of thought. I will gloss over many details here, as it's not relevant to the overall argument I'm making, but roughly, Bayesian inferential reasoning works like this: When trying to determine the likelihood of something, our minds go through a probabilistic syllogism of sorts. For instance, if I look outside and see wet concrete and wonder, *Did it recently rain?*, my brain measures the likelihood of rain generally with some "posterior" data such as how often I have seen wet concrete immediately following rain, what region of the world I am currently in, and so forth. These data are referred to as "priors" by many theorists. All of this gets tested against my background knowledge about the probability of wet concrete (1) as a result of rain versus (2) not as a result of rain. Then I make a "best guess" as to the cause of the wet concrete. Most if not all of our thinking, it is often argued, occurs in this manner. However, it is important to note that most of this inferential reasoning is happening below the conscious radar. To be sure, we often engage in purposeful or higher-order predictions. I plan my next vacation and use my "priors" to guide my assumptions about what is best, which might include things like my experiences flying at certain times of the day, the weather in the Southern Hemisphere, and so forth. But a lot of the predictive strategies we employ, Clark argues, end up being "non-conscious guessing that occurs as part of the complex neural processing routines that underpin and unify perception and action" (2016, 2).

Perception, in this view, is not prior to nor the condition for any probabilistic inferences, but is instead the result of these guesses. Instead of a passive observer bombarded by all manner of sensory signals, the perceiver is engaged in a world-making by inferring what is most likely the case, and all of this is the prerequisite for sensing any one *thing*.[3] Consider this simple example: you gaze out your office window and see the library building across the quad. You casually wonder whether a book you ordered has been delivered to the front desk yet, and this leads you to speculate whether the person at the front desk with whom you typically chat still works there because you haven't seen her in a while. It certainly *seems* as though you first see the building and then begin to analyze the scene into its composite parts—the inside of the building, the people there, the potential book waiting for you. But think about how the perception of the building *as a building* occurred in the first place. You don't actually see a whole building. Your visual system is only aware of the small slice of one side of the building that is available to you from your vantage point in your office. And yet you perceive the object as an entire building. Alva Noë (2009) refers to this feature of perception as its "presabsence"—that objects are always partially present and partially absent to us, despite the fact that we see them as whole. The posterior data available to you is highly convincing here. How many times have you looked out of a window upon a building-like structure only to find out, upon walking toward that building, that it is only a façade, with no depth and no insides? Your inference that, indeed, you are gazing upon a "normal" building is itself based upon a large data pool of normal building experiences. As Dennett (1991) has suggested, the scene "suggests" the notion of a whole building to you, and your brain takes the suggestion. But of course your brain is not just passively accepting the suggestion—it is testing it against all the priors and then settling on the most likely story.

In short, predictive processing (PP) asserts that a great deal of cognition—from the bottom-level neural activity that gives rise to more complex perceptual models, to the metacognitive theorizing that in turn exerts a top-down influence on the very basic processes—is predictive. Even perception, as we just discussed, is a sort of reasoning from uncertain or incomplete data. Likewise, imagination, according to Clark, might be a sort of virtual (PP whereby we are conducting thought experiments, using the past to guide our thoughts about possible future events.[4] Of course, not every system will yield the same level of information. Rudimentary neuronal sequences that are geared specifically to tracking incoming signals from other parts of the body almost certainly do not model and construct the world with anywhere near the same richness as those patterns that engage in determining precisely where to throw a disc such that the dog can run and catch it. Nevertheless, at every level, whether conscious or not, PP asserts that it is guessing *all the way down*. And for Clark, this guessing "provides the common currency that binds perception, action, emotion, and the exploitation of environmental structure into a functional whole" (2016, 4).

I have provided a condensed version of what is assuredly a detailed and compelling argument that Clark gives in *Surfing Uncertainty* because I am only interested in the overall gist of PP as it applies to the human-dog dyads we have been discussing. It may seem a far stretch to connect what I have argued regarding thinking-with dogs and this predictive account of cognition, but the link is not as convoluted as one might assume. First, Clark himself already does some of this work in a final section of the book, "Predicting with Others," where he notes that other agents' actions are also the stuff of prediction. We predict what others will do and how to act with them using the same generative model we use for other forms of prediction. Other agents, however, are also predictors, and so the story gets complicated, especially when we consider how we might go about minimizing error when interacting with these predicting agents. Clark thinks we are bound up in a process of "continual reciprocal prediction," much like his earlier work (1997) where he emphasized the continual reciprocal causation between humans and tools. These ideas dovetail nicely with Mitchell's account of "escalating reciprocity" as it occurs between humans and dogs. Before making that connection, I want to first consider a few potential problems with Clark's PP model, so as not to uncritically assume its merits among a host of criticisms, especially from the realm of enactivism, which I have been so ardently supporting thus far.

The first worry we might have regarding PP is a matter of limitation. Clark is clear that the account he is giving is for minds "like ours," and his focus is decidedly human. It might not be the case that dogs, or any other nonhuman animal, can be explicated in terms of prediction error, predictive coding, Bayesian inference, and so on. Again, the charge of uncritical anthropomorphism looms large here. But I think there is an easy response to this concern. If we consider the "ant dance" that Cheng discusses, it is not unreasonable to assume that the saccading the ants are doing serves predictions they are making about where best to move next, and furthermore, as they act in accordance with those perceptual cues, their predictive capacities are enhanced. In other words, the ants are engaged in what Clark calls "multilevel learning" (19), even if they are not consciously aware that they are learning. Recall that predictive processing can take place at an unconscious, nonlinguistic, and nonrepresentational level. Moreover, the story Clark tells in *Surfing Uncertainty* fits squarely with the idea that a lot of these predictive strategies humans deploy that are complex, and perhaps beyond what many nonhuman animals are capable of, are bootstrapped from the simpler ones.[5] Given the neural similitude between parts of the human brain co-opted in these simpler predictions and nonhuman animals' brains, it stands to reason that a similar PP account could apply to many species, albeit in varying degrees of complexity. This is not to say that human brains have somehow evolved more rapidly than nonhuman brains or that humans represent the most sophisticated by-product of predictive evolution. That is a contentious debate that is irrelevant to the discussion here. It is not contentious to say that given the context of the interactions humans have had over the millennia, as opposed to creatures like ants, the strategies we have developed reflect those environmental, sociological, and physiological differences. Dogs, while physiologically distinct in many important ways from humans, have shared environmental and sociological niches with us for thousands of years, and so we should expect parallels in at least some of the predictions they make within those niches. In some sense, this whole book so far has been an attempt to demonstrate what those predictive parallels are, and it was in play that we found the richest source of overlap. I will return to predictive processing in play after addressing what I think is an even more critical concern for PP, namely, its supposedly being at odds with an enactivist account.

Clark has been defending an *extended* view of cognition for quite some time now, so the PP model seems prima facie to adopt the standard cognitivist framework of describing the mind as a symbol-crunching computer that is housed neatly between the ears. This is not necessarily a bad thing—philosophers can and indeed ought to change their views if the evidence is overwhelmingly in favor of an alternative argument. And to some extent, that is what Clark has done over the years. In fact, even in *Supersizing the Mind* (2008) he argues for a more *organism-centered* account of cognition, such that while cognitive processes can and do extend beyond the brain, they always involve the brain as a central component. In *Surfing Uncertainty*, Clark seems to abandon the language of cognitive extension altogether, instead opting for more enactivist terminology, as can be seen in the quotes I have included thus far. He even suggests that multilevel learning is a form of self-organization, or what enactivists term *autopoiesis*, and that these autopoietic systems are marked by "the capacity to engage . . . in ongoing cycles in which perception and action work together to quash high-precision prediction error" (271). Rather than thinking of environmental features as *parts* of the cognitive processes, therefore, they are now seen as vehicles of prediction. I quote Clark again: "Learning is now grounded because distal causes are uncovered only as a means of predicting the play of sensory data (a play that also reflects the organism's own actions and interventions upon the world). Such learning is structure-revealing, unearthing complex patterns of interdependencies among causes operating at different scales of space and time. All this provides a kind of palette of predictive routines that can be combined in novel ways to deal with new situations" (2016, 270). Here we see that not only is Clark stepping more onto the enactivist side of the fence, but he is also indirectly addressing a long-standing objection to his hypothesis of extended cognition, namely, the *coupling-constitution fallacy*.[6] The objection is simply that just because some feature of the environment aids in the cognitive process, this does not mean it is a part of that process. When I am using my phone's GPS to find my way to the restaurant, and I form the belief that I should *turn left here*, this belief does not comprise my brain plus my phone. We are coupled to all sorts of environmental and sociological "aids" in this way, but this does not entail that they become a part of our minds. Clark redefines those "distal causes" as mechanisms that drive the predictive strategizing. There is still a worry that if cognition *just is* predictive processing and if environmental features are means of predicting, an avid externalist might argue that those features, so long as they are driving the predictive processing, are in fact part of the cognitive process. We can set aside this discussion for now, as it gets us too far afield, but suffice it to note that I remain mostly unconvinced of a difference that makes a difference between the hypothesis of extended cognition and some versions of enactivism when it comes to differentiating between a cause and a constituent.

What matters most here is the problem of Clark identifying PP within an enactivist framework while still relying so much on the brain as the "star player" in the cognition game. Indeed, enactivists such as Di Paolo et al. (2010) are often quick to dismiss the idea that the brain is any more important than the rest of the body or the world in co-constituting thoughtful action. While I agree that the overemphasis on the brain might be problematic, at least to many enactivists, I don't think that PP is completely at odds with enactivism, and especially not with the sort I have been defending in this book. In a 2016 paper, Gallagher and Allen attempt to show that a predictive model of cognition need not be at odds with enactivism. They propose a slightly amended version of Clark's PP model—what they call *predictive engagement* (PE)—to avoid falling into an overly internalist *predictive coding* (PC) model. A lot of PC models stress the symbol-crunching, computational, and Bayesian nature of predictive inference, which, as we discussed above, would locate these predictive (that is, cognitive) processes entirely in the head and thus detached from the niche in which they occur. PE, on the other hand, utilizes Clark's PP model but adds to it the idea of a "hermeneutic situation" to explain how predictive inference turns out to be an "active inference" in which an organism is inseparable from its situation. This is especially the case, they argue, in intersubjective contexts where social cognition is deployed.

Gallagher and Allen begin by noting that for enactivists, autopoiesis is not just about minimizing error and making correct predictions; autopoiesis is more about action than inference. "It's a doing, an enactive adjustment, a worldly engagement—with anticipatory and corrective aspects already included" (2016, 8). This attunement to the world encompasses the whole system, and in active world-engagement that system may very well turn out to be much more than a brain within a body. Nevertheless, another crucial feature of autopoiesis is to maintain homeostasis so as to avoid entropy, which is to say death. There is, in other words, a centralized system to which these engagements in the world matter. Thus, while predictive processing might just be one among several built-in features of self-organization, it is an important one and likely accounts for a great deal of our transactions in the world. The loop between perception and action, Gallagher and Allen note, "provides a deeply embodied form of engagement, where the priors and actions an organism is likely to entertain are fundamentally constrained and afforded by the morphological structure of the agent's body. In this way, the Bayesian brain is uniquely equipped to exploit the finely tuned properties of an organism's dynamic morphological body and associated Umwelt" (10). This idea resonates with much of what Clark argues regarding the "loops" that serve to tune and refine human thinking, such as sketch pads, smartphones, and even other humans.[7] These tools show up as meaningful to humans because of a particular umwelt to which humans are accustomed, and, as Gallagher and Allen rightly point out, the actions we are likely to entertain with these features of our environment are always already constrained by the types of bodies we have. I cannot, for example, expect to read a road sign in the distance if I do not have my glasses on. And if I am trying to survive, something I most assuredly do when driving, and hence why I should always wear my glasses, I will need to be able to read road signs to predict with minimal error when to exit the highway, what speed to go, and so forth.

Gallagher and Allen further attempt to reconcile predictive processing with an enactivist account by arguing that there is a hermeneutic element to the organism-world engagement. Drawing on Dewey's (1938) and Gadamer's (1989) accounts, they note that in the hermeneutic tradition, one cannot separate out a situation from an agent. For Dewey, it is incorrect to equate a situation with an environment, or at least it is wrong to think of an environment as simply the physical location of a situation. A situation might be something like playing keep-away with my dog in the backyard. While it might be tempting to think of the environment here as "the backyard," the players of the game as myself and my dog, and then the game itself, a hermeneutic interpretation would see these as inextricably bound up in the situation of "playing a game." The whole process of play, in this sense, co-defines me, my dog, the space we are sharing—it's not just a backyard but a "playground"—and the interactions in which we are engaged. As Gallagher and Allen put it, "the agent cannot step outside of the situation without changing it. If I am in what Dewey calls a problematic situation, I cannot strictly point to the situation because my pointing is part of the situation. My movement is a movement of the situation—and a rearrangement of objects in the situation is a rearrangement of oneself as well" (12). Indeed, it makes no sense to think of the game as somehow over and above the players and their gamespace, but moreover, each adjustment and every move made in that situation changes not just the game but the players and their shared space.

Consider how the "hermeneutic situation" was employed by Mitchell (2015) and Mitchell and Thompson (1991) in their descriptions of creativity within human-dog play dyads. Although they did not use this terminology themselves, it is present in their discussions, especially when linking creative human-dog play with creativity in sports. The way boxers, for example, engage in the collaborative practice of "practical improvisation" parallels the argument Gallagher and Allen give in several ways. The boxers' movements serve to guide the predictive engagement, much as in the dynamic projects observed between humans and dogs at play. We cannot hope to decompose the "situation" of this play into place or space, individual players, and individual movements. Furthermore, as each player in these games adjusts in response to the other's moves, the situation is transformed, and as these situational transformations unfold, we can see how the "rearrangement of oneself" occurs in each of the players. Think of how Chris attempted repeatedly to fake out Hercules, for instance. This routine slowly gave way to Hercules correcting his movements to match the attempted deception so he could "catch it" before it happened. His innovative strategies shaped the routine into something quite different, which in turn scaffolded Chris's movements.

Gallagher and Allen tie this discussion about the hermeneutics of situational engagement to how we think about social cognition, to argue that mindreading is really just an instance of *predictive engagement* within a social situation. When we enter into an exchange with another person—be it a conversation between adults or even just a parent-infant nonlinguistic interaction—we do so with all sorts of background information about our partner, the space, the context, and so forth. If I am at a conference dinner with colleagues and a guest speaker, I will already know quite a lot about my colleagues from working with them, and I will know at least something about the guest speaker from the talk just given. Furthermore, I know the norms and stereotypes of academic dining. These "priors," in the language of predictive coding or predictive processing, are not just static pieces of data I am crunching in my brain in order to determine what is most likely to occur at each moment of the dinner. They are dynamic and constantly changing humans who are themselves predictors of my actions. Likewise, all the social systems at play that serve to feed my predictions are ever changing, and they are also constituted by the agents with whom I'm interacting. Even in cases of infant-parent dyads, where the baby arguably has very little in terms of priors to go on, there is a sense in which the familiar matters here. This is seen when babies behave quite differently around strangers as opposed to a parent or familiar guardian (Brooker et al. 2013). Moreover, the interactions are precisely what drive the capacity to predict with others better in the future. As we discussed in previous chapters, affective co-attunement and mindreading are constituted by these very important infant-caregiver dyadic exchanges.

Social cognition, therefore, is an example of what Gallagher and Allen call an *enactivist hermeneutics*. Unlike *neural hermeneutics* (see Frith and Wentzer 2013; Barrett and Simmons 2015; Friston and Frith 2015), which locates the attempts at reaching understanding in the brain, enactivist hermeneutics argues that the brain is one of many components co-opted in the task of social know-how. The brain is shaped by social interactions, to be sure, just as brains are affected by embodied experiences. We know, for instance, that taxi drivers' brains are altered by the long experience of navigating city streets (Maguire et al. 2006) and that dancers and nondancers have differently structured brains (Calvo-Merino et al. 2005, 2006). But the brain is not the only thing transformed by these interactions. The agents—whether humans, dogs, or any other beings involved in a social exchange—alter *as a result of* the interaction. It might be true that very important bottom-up processes are occurring that make such transactions possible; as Clark rightly notes, a lot of these predictive processes are happening far below the conscious radar and are thereby affording much more sophisticated forms of predictive processing. However, those bottom-level processes are not what we, the subjects of experience, are engaged with. In dynamic social exchange, whether in communicating with a human partner or playing fetch with my dog, enactivist hermeneutics argues that we must carve the joints of agency where they occur phenomenologically, namely, with the *whole agent engaged in a situation with me*. Quoting Gallagher and Allen at length summarizes the view nicely:

> Just as the hand adjusts to the shape of the object to be grasped, so the brain adjusts to the circumstances of organism-environment. Rather than thinking of this as a kind of inference, enactivists think of it as a kind of dynamic adjustment process in which the brain, as part of and along with the larger organism, settles into the right kind of attunement with the environment. Social interaction thus involves the integration of brain processes into a complex mix of transactions that involve moving, gesturing, and engaging with the expressive bodies of others; bodies that incorporate artifacts, tools, and technologies, that are situated in various physical environments, and defined by diverse social roles and institutional practices. (2016, 17)

Gallagher and Allen correctly note that this distinction between predictive processing and predictive engagement is not merely semantic; indeed, despite some enactivist-sounding phraseology, Clark does irresistibly lean back toward a more brain-centered story regarding thought. However, the last line of the quote above does resonate with much of Clark's position regarding "designer environments" (2016, 275–81). For example, he argues that "action and perception then work together to reduce prediction error only against the more slowly evolving backdrop of a culturally distributed process that spawns a succession of practices ... whose impact on the development ... and unfolding of human thought can hardly be overestimated (280). The chief difference, of course, is that Gallagher and Allen, along with most enactivists, emphasize not just the "slowly evolving backdrop" of sociocultural systems but also the immediately available and ever-present ones that dynamically change and are changed by the agents enacting them.

So far, I've connected the PP model with a more radically enactive account of cognition that sees prediction more as active engagement. I've provided arguments as to how we not only think-with other persons but with other animals too, which can be seen perhaps most clearly in dynamic play between humans and dogs. This play can be understood as predictive engagement, which allows for the "hermeneutic situation" to both guide and be changed by the players. And these players, during bouts of play, are better thought of as essentially tied to one another, rather than as encapsulated and isolated organisms. That is, we first recognize the bout of play as a whole and then go on to analyze the "individual" players, their thoughts, actions, and so on. What is left is to tie all of this back into the discussion of creativity and innovation as these elements occur in play and other modes of thinking-with. Indeed, one of the central thrusts of this chapter is to more fully convince the reader that rather than *autopoiesis*, what enactivists ought to be arguing is going on in social cognition is *sympoiesis*. It is not enough to stress the self-organizing and self-sustaining aspects of organisms when explicating how they think with others, because in thinking-with, we are always already *making-with*—making meaning, innovating new ways to communicate, and creating understanding. And when attempting to explain a complex phenomenon such as interspecies communication in playful dyads, I think we need to heed quite seriously Despret's charge that the unit of analysis for interpreting these situations is not two separate organisms in a shared environment but a multiorganismic system that emerges *in the interaction itself*, and in turn shapes future interactive possibilities.

I think Clark, perhaps unwittingly, provides an excellent example of sympoiesis, despite his insistence on brain-centrality, in the following example: "An idea which only Joe's experience makes available, but which can flourish and realize its full potential only in the intellectual niche currently provided by the brain of Mary, can now realize its full potential by journeying between those agents. Different agents (and the same agent at different times) constitute different 'filters,' and groups of such agents make available trajectories of learning and discovery that no single agent could comprehend" (2016, 288). Rather than "in the brain of Mary," an enactivist hermeneutics or sympoietic account of this exchange would argue that Joe's ideas attain their full intellectual potential against the backdrop of Mary's full array of background knowledge. They both bring to the table embodied know-how, but also *know-with*, assuming Mary and Joe are familiar and share familiar modes of typical communicative exchange, along with all the prior social conditions that make such an exchange possible and meaningful in the first place. This is clearly not just *thinking-with*, nor can autopoiesis capture the full extent of the interaction and what unfolds within it. Mary and Joe are engaged in a creative socially cognitive act that cannot be understood by any of its requisite parts alone. In this making-with, the unit doing the creating is Joe-with-Mary, and they are both shaping and being shaped by the interaction.

Note that in this description of Joe-with-Mary we have the slow, distal and evolutionary causes at play serving to shape the interaction—human linguistic evolution, gestures, all of the ideas and innovations leading up to this creative moment—but the here-and-now dynamics of the conversation also undoubtedly play a constitutive role. Verbal and nonverbal confirmations, questions, redirections, and arguments push the ideas around and transform them, the way two boxers literally push each other around, thereby transforming each other and the game itself.

This is precisely the story I have been telling about humans and dogs. Thousands of years of shared history, embodied thinking-with, and sociocultural exchanges have afforded the two species a vast backdrop of communicative potential. When we observe humans and dogs interacting in the here and now, all of those priors, we might say, are the conditions for the possibility of the meaningful exchanges taking place. They are causes, but not necessarily constituents, in other words. However, the dynamic gesturing, developing projects, and affective attuning that are occurring online in real time are not just causes of the interaction; *they are the interactive moments themselves*. Thus, if we are to understand how humans and dogs can be cocreative, in play or in any other exchange, I argue it is time we focus on humans-with-dogs more closely. That is, instead of perseverating on the vast differences between the species or trying desperately to find commonalities among them, I am suggesting we treat them more like *siphonophores*, which are the ideal sympoietic systems. These are creatures, such as the Portuguese man-of-war, that are composed of many independent organisms who attach themselves to one another so as to be functionally complete (see Fewkes 1880; Haddock et al. 2005). The man-of-war, for instance, is often taken to be a singular jellyfish, when in fact it comprises many singular zooids that have come together because they each lack capacities that the others have. The result is a quite powerful and venomous ocean predator—or, more correctly, collective of minipredators. Siphonophores, like holobionts, are useful analogues for how we ought to conceive of human-dog pairings.

Of course, I am suggesting this connection between siphonophores and dogs-with-humans only as a theoretical analogy, and only for the purposes of understanding tightly coupled cases of making-with, such as those occurring in collaborative play or certain affective exchanges. Dogs and humans can and do "decouple," and accordingly can be explained as single organisms, capable of autopoietic functioning as well. Nevertheless, in those situations—or, as Despret calls them, "apparatuses"—under which shared creativity defines the interaction, siphonophores are an apt analogue.

But there is an important difference besides the potential to decouple, and that is that humans and dogs also arguably *intend* to interact and play together. As Burghardt argues regarding the condition for play, it only counts if the creatures involved are not doing it to survive or because it is stereotyped behavior. Siphonophores are likely hardwired to do what they do. They are not creatively making-with, even if they do "make up" something. This difference is how we can respond to the worries raised earlier about characterizing *all* forms of environmental engagement as cognitive, from the bacterium's interactions with ant bodies to the human's interaction with a legal system. Humans and dogs playing together is *siphonophoric* insofar as it materially takes shape in a similar way to a man-of-war, but it goes beyond that. The players in *this game* are aware of the game and of one another, gauge the others' next moves, and plan appropriate responses. And all of this happens in a dynamically unfolding way that is both creative and predictable.

This argument for sympoietic creativity in human-dog dyads would not be very convincing without returning to the issue of predictive engagement, specifically as it concerns *breakdown*, or misunderstanding. It is not always the case that humans and dogs perfectly well understand one another. In fact, as we have observed, it is often the case that dogs seem to "get us" better than we make sense of them. Thus, in the final section of this chapter, I want to examine how it could even be possible to conceive of a sympoietic system in terms of predictive engagement. If, as I have been arguing, the unit of analysis in certain interactive situations ought to be human-with-dog, then how might we say that one is attempting to predict the other, or that one has *failed* to predict the other? In order to answer these questions, I want to take a closer look at what Cheng (2018) has referred to as the "scribbling and bibbling" that takes place when an organism embedded in its umwelt—especially in a social umwelt—makes predictions, acts, and engages in meaning-making.

Scribbling, Bibbling, and Enacting Collaborative Cognition

I brought up siphonophores primarily as a useful analogue to the human-dog dyad because in certain contexts, such as in playful collaborations, humans and dogs are so tightly conjoined that their actions, reactions, and cognitive processes are bound up in a unified whole. This larger, multiorganism unit—the human-with-dog—is not perfectly analogous to the siphonophore, of course. Dog-with-human is more of a pragmatic notion, one that serves as a useful way to analyze how meaning, affect, and attunement emerge in dyadic exchange. Dogs and humans can "decouple" and go about thinking and feeling without one another, much in the same way that a human can put down her smartphone and attempt to navigate without the use of GPS. In the case of human-dog coupling, we still have two individual species that can and do live, think, and act on their own, whereas a siphonophore, by definition, must include all of its "parts"—that is, organisms—to function as a whole. Nevertheless, there are many projects that humans and dogs undertake that are made much more successful by a tight coupling that resembles in many ways how a colonial animal like the man-of-war functions. Human-dog search-and-rescue and cadaver teams are an excellent example of how each participant in the dyad serves a unique function, resulting in outcomes that far surpass how the individuals, human or dog, would fare on their own.[8]

In his 2018 paper, Cheng argues that *Cnidaria*—the phylum to which siphonophores belong—might well be considered exemplars of embodied cognition. Even though they lack a centralized brain, the members of *Cnidaria* all exhibit varying forms of thoughtful action. Anemones and hydra learn via classical conditioning and habituation (Rushforth et al. 1964; Haralson et al. 1975; Logan 1975; Logan and Beck 1978; Johnson and Wuensch 1994), and anemones fight with one another (Ayre and Grosberg 1995, 1996). Siphonophores engage in carefully articulated hunting patterns and exhibit mindedness in their planning and execution of concerted actions (Albert 2011). These creatures all show signs of predictive engagement (PE), in other words. Using tentacles and cilia, they "sample" areas for potential food and navigate based on biofeedback. In the case of siphonophores, Haddock et al. (2005) have found that *Erenna*, a species of deep-sea siphonophore that prey primarily on fish rather than on crustaceans like most members of the phylum, use bioluminescence to lure the fish to their venomous nematocysts. Although bioluminescence is widely observed in sea life, it is most commonly seen as a defensive strategy and rarely as a lure. What makes the discovery even more compelling is that the light emitted by *Erenna* is red, with the longest wavelengths, attracting fish who otherwise would not notice the light at such great depths. To be sure, much of this luring behavior is likely nonconscious and probably the result of a long coevolutionary process between the siphonophores and the visual capacities of the fish upon which they prey. But recall that for PE it is not required that the animal know it is making these predictions; the predictive engagement simply must drive along the action-perception loop. Thus the luring behaviors of *Erenna*, along with the other forms of predictive engagement displayed by species among *Cnidaria*, are convincing cases for the enactivist argument that a centralized brain is neither necessary nor sufficient for thoughtful action. As Cheng argues of *Cnidaria*, "the neuroanatomy suggests that embodied cognition might well rule the day" (2018, 12).

Again, siphonophores are intriguing because not only do they possess the capacity to predictively engage in the world without a centralized brain or nervous system, they are quite literally doing this enactive cognizing with others, because, after all, a single siphonophore is, rather paradoxically, many creatures. All the organisms that make up the siphonophore have dedicated tasks—the individual called the necto-calyx, for instance, is the "swimming bell" and propels the whole system through the water, while the tentacular knobs "see by feeling" in order to determine where to go and where to avoid—and yet they all work in concert, connected only by a very long axon or stem. Each individual communicates the feedback it receives from its environmental or interoceptive sampling to the whole colony, and meaningful action emerges. By "meaningful action" I refer both to the here-and-now individually intelligent behaviors such as luring and escaping from predators and to the more collective actions of the whole order of siphonophores, a group that have surely figured out a successful survival strategy. By altering and diversifying their behavior and physiology over the course of 500 million years, siphonophores are among the most prolific animals.

It is worth noting that the thinking-with/making-with story I am telling here has so far been cast in a mostly positive light. Indeed, siphonophores have figured out how to communicate well, which is both a predictor and a result of successful engagements within their umwelt. But as Clark (2016) points out regarding his own PP model of cognizing, predictive strategies aren't fail proof. Likewise, prediction will not be perfectly accurate within a sympoietic system, whether intra- or interspecies. Clark rightly notes that we predict with other humans in many ways, but one of the most prevalent ways we understand each other, plan with and around each other, and innovate together is through our use of language. Of course, even in linguistic exchange, breakdowns can and do occur. Although my aim here is to discuss nonlinguistic interspecies interactions between dogs and humans, I want to briefly consider human communication and miscommunication as it occurs in language, since this phenomenon provides a useful analogue to the human-dog dyadic exchanges that are the focal point of this book.

In a recent paper, Gallagher (2020) discusses "intersubjective alignment" as it occurs in conversational and embodied contexts. The idea parallels what I have been calling "attunement," coordinating actions, postures, and attitudes to facilitate an interaction. Alignment occurs when dynamic interconnections are formed that shift and change as the exchange unfolds; hence, a conversation makes an apt example of such an "attuning" process. Conversations are almost always "unchoreographed," although they may begin with a specific plan or topic in mind. I might, for example, pop into my colleague's office intending to determine which of us will lead an upcoming committee task. Once the conversation begins, however, it takes on a life of its own, and a point my colleague raises about why I am better suited to leading might spark a memory of a past situation in which my leadership was effective. In turn, my response could cause a reconsideration of something my colleague said earlier, which might result in a small digression about an unrelated matter. In short, as Gallagher notes, simply because we cannot perfectly predict what our interlocutor will say next, conversations have a decidedly spontaneous element to them. All the same, they are "aligned" insofar as there are constraints at work. I might not be able to predict with 100 percent accuracy what my colleague will say, but I have a good idea of what will *not* be said. Discussing our committee tasks will most likely not elicit a random interjection about our favorite ice cream flavors. There are some regularities and synchronicity to our conversation, in other words. All of this is because, as Gallagher argues, in alignment the interconnections formed are "dynamic and adaptive to environmental constraints as well as to higher-order cognitive processes such as individual and shared or collective intentions, and at the same time they generate ongoing, shared cognitive states" (2020, 66).

Another argument Gallagher makes regarding alignment is that it is wrong to think of it as either top-down or bottom-up. Rather, he claims, we ought to conceive of these interconnections as forming in a gestalt manner, such that "higher" level factors, such as cultural norms and social institutions, and "lower" factors, such as basic sensorimotor contingencies and neurobiological processes, are in a constant figure-ground relationship, mutually and simultaneously informing and being informed by the other. In the same way, he argues, we ought to think of "memory or imagination, not as separate processes from conversational or interactional practices, but as continuous with and intervening in speech, gesture, posture, movement and action" (2020, 67). This idea dovetails nicely with what Clark (2016) argues about how predictive processing is supposed to work in a dynamic and interconnected way, from the bottom to the top and back down again. Of course, Gallagher is going to favor referring to any predictive inferences made in these exchanges as *predictive engagements*, and as I have said, I tend to agree with him. The idea that all the subpersonal and personal processes, as well as the interpersonal ones, are occurring in dynamic tandem loops is nonetheless similar between Clark's and Gallagher's accounts.

Conversations, as they emerge from the back-and-forth between two agents, are paradigm cases of *joint actions*. Joint action involves material engagement in dynamic contexts. These actions can be explicated in terms of differences in cultural background or "habitus." Such things in fact characterize many everyday actions and practices, as well as rituals within institutional contexts (Gallagher 2013; Gallagher and Ransom 2016). Conversations, described in this way, are a lot like play. While we have seen there is good reason to think of play in *teleofunctional* ways—namely, that there is a purpose to play, and that intentional or goal-directed actions are what comprise each project formation, augmentation, or transformation—play is also quite spontaneous, especially if it is "free" play, not a game with detailed rules or regulations. Much as in conversational exchange, there can be miscommunication or breakdown in playful dyads. If I am caught up in a game of fetch with my dog, and she suddenly drops the toy she has been fetching in a nearby bush and then bows at me, I might interpret this as an enticement to come over and try to get the toy out of the bush. She might decide to snatch it back up just before I get to the bush and initiate a game of keep-away. Or I might get close to her, assuming she will initiate this keep-away game, only to see her dart away without the toy, chasing the cat instead. In all play, but especially in interspecies play like this, we often misread each other and guess incorrectly about the next move. As in conversational breakdowns, however, all is not lost just because there is a moment of confusion. We seek clarification from our conversational partner, in the same way I might try to understand my dog's actions as she wavers from what was expected. I cannot ask her to explain herself, not in words, but I can entreat her to tell me what she means all the same. I can follow her around the yard, wait until she has gotten the cat-chasing fever out of her system, or, perhaps more cunningly, "play pretend" that I am not interested in her sudden detour from our game and see if my disinterest causes her to regain interest.

In cases of communicative breakdown—whether in conversations, in play, or in any other dyadic exchange—we could provide an explanation for the correcting of prediction errors that closely mirrors Clark's account of PP. Namely, the signals we receive do not match up with what was predicted and thus our brains have to readjust, recalibrate, and repredict, based on the new data. This account, as I have argued, is far too brain based, subjective, and internal. The model of PE that Gallagher and Allen defend, on the other hand, allows for these "recalculations" to occur intersubjectively. Or, as Gallagher's terminology might suggest, when trying to make sense of an interaction that has seemingly gone wrong or broken down in some way, I need to "realign" myself with the other. In order to do this, it won't suffice for me to retreat into my own head and run some simulations. Instead I must continue to interact, or, in the language of prediction, I need to resample.

Cheng (2018) calls these adjustments and corrections the "scribbling and bibbling" of cognition. While he once argued that much of this scribbling and bibbling occurs "inside the head" in a classically cognitivist, representational manner (1995), he has since changed his views so as to emphasize the action involved. When discussing navigation in ants, and then navigation more generally, he argues that the process is a prime example of how an enactivist framework better accounts for what is actually happening. If we are trying to explain how an animal is engaging in the cognitive process of navigating through its world, we must do so teleofunctionally. Just as we cannot separate the play activity from the playing dyad, and much as Gallagher and Allen (2016) characterize hermeneutic situations, to speak about navigation as a purely representational and internal process would be to miss the point entirely. Navigation is a process of interacting with the world with the purpose of *getting somewhere*. In his 1995 article, Cheng described navigation as *servomechanistic* in that the process was largely representational—a "comparator" measures readings of variables against what those variables ought to be, while the error or discrepancy would then drive the action. This view of a servomechanism underwriting navigation shares some similarities with Clark's 2016 PP model, because, as we saw, even though Clark is clear that actions are part of the loops that feed prediction, the predictive inferences themselves occur in the brain. Cheng's more recent views keep the idea of a servomechanism but de-emphasize the representational component to show how action is just as deeply entwined in the process of navigation. His argument is that noncognitive actions play a causal role in cognition, not necessarily a constitutive one. However, I think his description of the way ants "pirouette" in seemingly choreographed fashion actually underscores the way that the actions *just are* part of the cognitive process itself, thereby buttressing the bolder enactivist claim that thought and action cannot be disentangled.

Finally, in the case of intersubjective collaborations, not only should we be emphasizing the actions—or, better, the interactions—but those interactive moments *just are* the cognitive processes themselves. To echo Noë's 2009 idea once more, that thinking is a sort of performance, in sympoietic *thinking-with* these collaborative dyads cocreate and perform thought-in-action. The collaboration itself is where the meaning and significance are to be found. There is no internal intent or subjective theorizing to be decoded from observable behavior, as the behavior is just one part of the interactive moment, the whole of which constitutes the thinking itself. As we have seen, human-canine dyads are often quintessential cases of *sympoiesis at play*. Dogs-with-humans can and do form collaborative dyads that in turn create coactive cognition. This cognition cannot be understood by studying either member of the pair in isolation, nor can standard cognitive science do justice to the complexities of the interactions. In many instances, dogs and humans constitute a tightly conjoined "superspecies" that the traditional brain-based models in philosophy and science will continually fail to understand. The "radical" idea that cognitive processes are often bigger than what is to be found within a single organism's brain or body is not so radical after all. The future of cognitive science should be *collaborative*.

POSTSCRIPT

Six-Million-Dollar Dogs and the Future of Companion Species Studies

During the 2020 Super Bowl,[1] a commercial aired in which a dog named Scout is playing on a beach. A human voice narrates as if it were Scout telling his story, and we learn that Scout has beaten all odds by surviving brain cancer, despite a 1 percent chance to do so. We further learn that this is because of a cutting-edge research program at the University of Wisconsin–Madison, where Scout received his lifesaving treatment. The commercial is dripping with sentimental anthropomorphism, but it is effective *and affective*. Even the staunchest skeptic with regard to how animals might think like humans—what Andrews refers to as an "anthropectomist"—is likely to be moved by Scout's story.[2]

Anyone familiar with advertisements during the Super Bowl knows that they are costly. Scout's commercial cost a whopping $6 million, and it was paid for by Scout's guardian, David MacNeil, the founder and CEO of WeatherTech, a U.S.-based company specializing in motorized vehicle accessories. In an interview with NBC (Tibbles 2020), MacNeil says he paid for the advertisement to show thanks to the doctors at UW-Madison and to raise awareness of what they are doing. He also hoped to encourage donations to the School of Veterinary Medicine at the university, to facilitate even more lifesaving research and help more beloved animals receive care. The advertisement was successful—within minutes of airing, thousands of dollars began pouring into UW's program (Lucas 2020).

The intention behind MacNeil's efforts is undoubtedly good. As someone able to afford such an exorbitant commercial slot, choosing to focus on his dog's treatment and raising funds for the program that made it possible are indeed noble reasons. The story, however, also underscores some complex issues that can serve as a culminating point to this book. We began with a story of a human risking his life to save a puppy in rising floodwaters, and we end with a story about a human who not only spent $6 million on a commercial about his dog's cancer treatment but was also able to pay for the lifesaving treatment in the first place. There is no question that dogs are integral members of the family to so many humans around the world and that humans feel deep attachments to dogs, as well as all sorts of other nonhuman animals. Nevertheless, we might wonder to what extent such dramatic attempts to "rescue" companion species are justified. Treating brain tumors in dogs can range from $6,000 to $10,000 and is often not successful. In the case of radiation that might grant a dog with lymphoma another year or two of life, the procedures are around $5,000. According to Dr. Steve Suter, an oncologist at North Carolina State University, where he has performed more than seventy bone marrow transplants on cancer-stricken dogs, those cost patients between $16,000 and $25,000 (Berr 2015). Some people, like David MacNeil, simply pay for these procedures out of pocket. Less wealthy families might need to mortgage their homes, rely on the ever-burgeoning industry of pet insurance for assistance, or defray the costs with payment plans or medical credit cards. But the bulk of dog guardians are not in a position to spend $20,000 on cancer treatment. While some might turn to crowdfunding as a last resort (Young and Scheinberg 2017), the sad reality for many people who have a dog or other animal with cancer is that the animal is euthanized. Again, MacNeil's mission in airing his commercial will ideally be part of a larger effort to make treatments like Scout's more affordable in the long run, but as it stands now, it is a high mark of economic privilege to be able to provide cancer treatment to one's dog, or even to obtain the testing required to diagnose the disease in the first place.

Dogs are, in many ways, a reflection of human interest. When studied closely, they can reveal to us all sorts of philosophically, sociologically, and politically complicated facts about us and our implicit biases. Consider that in the United States, where universal health care remains to be an actualized and well-functioning system, humans who suffer from cancer often go bankrupt or turn to crowdfunding to afford their treatment (Cohen et al. 2019). Rather than making care affordable, either for humans or their nonhuman companions, we have exorbitantly high insurance premiums and impossibly expensive medications. Only the wealthiest and most privileged among us have genuine access to the full array of lifesaving and health-preserving services our society offers. Stories such as Scout's, while heartwarming in so many ways, also highlight the problematic inequalities inherent in our world. In the case of the U.S. health care system, dogs are showing us more and more the failures of capitalism.

All this discussion might appear far afield from what I've argued in this book, but I think the parallels are much stronger than one would expect. Consider again this quote from Haraway: "Bounded (or neoliberal) individualism amended by autopoiesis is not good enough figurally or scientifically; it misleads us down deadly paths" (2016, 33). Here she ties together science, philosophy, and politics and critiques them all for a stubborn insistence on individualism. Her charge that the addition of autopoietic talk will not suffice and might even be "dangerous" is echoed throughout my arguments regarding sympoiesis and coactive thinking and making-with. In this book I have offered a defense only of the first part of her claim, namely, that autopoietic enactivism falls short in explaining human-dog cognition. However, the second part, where she claims it might even be "deadly" to fail to properly account for what she calls "strange kin," is not so hard to envisage as following from my arguments. In other words, there is a deep connection between how we understand individuals, how those individuals might think, whether we afford those individuals subjectivity or rationality, and our moral obligation to them.

So I close with some points for consideration regarding these connections, between how we study dogs and other nonhuman animals and how we treat them, and in turn how these interconnected practices reflect and shape how we treat other humans. First, as I have argued throughout the book, conceiving of cognition among species as coactive and sympoietic will alter the way philosophers and cognitive scientists conduct their research. For instance, in philosophy, while problematizing individuals and part-whole relationships has often been viewed as the stuff of abstract and obtuse metaphysics, when we consider how very real and concrete these problems are in the biological world, there is less reason to relegate them to such obscure domains. In fact, with growing recognition of holobionts and superorganisms, it seems that if philosophers *don't* take the problematic assumptions regarding individuals seriously, they are overlooking important parts of reality and doing a disservice to their own discipline.

Philosophical reframing, on the other hand, can lead the charge in rethinking how things work in the sciences. If enactivism is generally accepted and we begin to view cognition as more of a shared and distributed process that is to be found on a continuum, where differences among species amount more to differences in degree than in kind, then animal researchers might be better primed to see cognition occurring in a more multimodal way. A view such as this would not be overly concerned with finding "humanlike" cognition in the nonhuman world, so much as with locating instances that demonstrate how animals are capable of cooperation, co-making, and co-attuning in their unique lifeworlds. And when those lifeworlds are inextricably tied to other organisms, as they so often turn out to be, we are better equipped to make sense of the symbiotic and sympoietic nature of such pairings. This applies to human-dog dyads, to be sure, but extends to all sorts of other "holobiontic" or "siphonophoric" systems.

For instance, when Sony's AIBO—a $3,000 robot dog—was being developed, researchers like Horowitz and Bekoff (2007) were interested in the programming that guided the artificially intelligent dog's actions. They went on to study how anthropomorphic ascriptions arise when humans play with their dogs and to what extent those tendencies were elicited by the same sorts of behaviors programmed into AIBO. While this research is intriguing in itself, AIBO also sparked interest among animal behaviorists including Ellen Mahurin, who was curious how living dogs would "get along with" a robot companion (Wollerton 2019). So far, the reviews are mixed, she claims, as many dogs seem uninterested in AIBO, or even scared of it. The most interesting question in all this, however, is *To what extent do the dogs see AIBO as another dog?* Mahurin argues the dogs she has studied overwhelmingly do not treat AIBO as a living dog. Her reasoning is based on behaviors the dogs exhibit or fail to exhibit while interacting with the robot. For example, they do not entice AIBO to play, nor do they become possessive of their toys or food when in AIBO's presence, something they would be prone to do around living dogs. They do, however, tend to sniff at AIBO's anal-genital area, which suggests that perhaps on some perceptual level, AIBO shows up *as an animal*.[3] We might say that the dogs then test out predictive strategies—they engage in sniffing where they would expect to glean important information, but then find none of the usual data points and so, like any other predictive engine, they alter their course based on this input. Notice that framing the dogs' behaviors toward AIBO in this way does not unnecessarily anthropomorphize them. Indeed, even ants who "scribble and bibble" about, as Cheng (2018) characterizes them, are likely doing a similar predictive "game," one that is specific to the ants' umwelt and the interactions therein. The predictive processing model discussed in chapter 5, therefore, can explain quite a lot about animal behaviors. But what causes dogs to "cynomorphize"? What would it take, in other words, not just for a human to accept AIBO as "dog enough," but for dogs to extend the same recognition? Are dogs susceptible to an "uncanny valley" effect like humans?[4] Questions like these, I argue, can only be properly addressed by considering the shared cognitive lifeworlds of humans and dogs, and furthermore by examining how technology shapes and is shaped by both species.

Placing technology like AIBO in homes with dogs inevitably raises animal welfare questions as well. It might be, for example, that the noises and movements AIBO makes are so frightening to some dogs that it would be considered unethical to subject our canine companions to it. The fact that we even worry about such concerns implies that the relationships many humans maintain with dogs are far deeper than common discourse might lead one to believe. In other words, dogs are not *pets* so much as they are *family members*. Throughout the book, I have subtly made this argument, resisting talking about dogs as pets, or humans as owners, but instead as companions and guardians respectively. This is not just a rhetorical flourish. If it is true, as Haraway argues, that one of the ways we engage in sympoiesis is by *making kin*, then we have a very real responsibility to carefully curate how we talk about animals. As ethological narratives get repeated, they become sedimented into culture. They shape how we think about animals and, in turn, ourselves. Dogs qua "pets" might connote subservience or a species far less intelligent than humans, and not a creature capable of genuine cognitive and emotional interspecies connection. Dogs qua "companions," on the other hand, can open a space for rethinking this relationship and, in turn, our role in it.[5] Rather than owners of pieces of property, we are partners with a materially different yet wholly familiar *other*, and we have an obligation to treat dogs with love and care.

This shift in thinking about our relationship to animals is of course well under way, and it is not limited to dogs. The country of India officially declared dolphins "nonhuman persons," and its Ministry of Environment and Forests recommends all entertainment-based aquariums that capture and confine dolphins or other cetaceans be banned (S. Coelho 2013). Switzerland recently joined several other countries in demanding that lobsters be humanely killed before they are cooked. Legislators cite research on animal pain and sentience showing that while it might not look anything like pain in humans, evidence points overwhelmingly to the fact that lobsters and other crustaceans can and do suffer when boiled alive (BBC News 2018). Philosophy of animal cognition, which informs and is informed by animal science, ultimately influences public policy and legislation.

However, as humans become increasingly aware of their duties to other animals in general, and particularly their deep bonds with and moral obligations to dogs, all sorts of new problems begin to arise. As I claimed above, dogs often illuminate failures on the part of humans, such as our shortsightedness in understanding the interdependencies of systems that shape and delimit how we think about highly complex issues. To get a sense of how myopia persists despite ever-increasing knowledge about animals and our relationship to them, consider the rise of emotional support animals (ESAs). As research demonstrating how affectively co-attuned animals—especially dogs—can be with humans became widespread, it did not take long for the general public to adopt an idea that had until recently been a bit on the fringe: namely, that dogs can be *therapeutic* (Froling 2001, 2009). They can calm anxiety, lessen depression, promote a healthier lifestyle, and assist with those suffering from post-traumatic stress disorder (PTSD). This recognition in turn shaped legislation, such that the Fair Housing Act now includes ESAs as "reasonable accommodations," allowing that these "assistance animals" be permitted in places of residence, including hotels, and on major transport vehicles, like planes and trains. It is interesting to note that ESAs, like service dogs and therapy animals, are specifically *not* considered "pets," according to this amended law. This gets ESAs around the no-pets policies so often found in public places. It also underscores the idea that changing how we talk about animals can and does have real-world consequences.

While there are obvious benefits to allowing animals in more places, not least of which is providing genuine help to people suffering from mental health problems, to assume that ESAs are an unqualified good would be an example of the shortsightedness I just alluded to. One need only conduct a bit of "Google research" to find countless examples of how the ESA boom has encountered stumbling blocks, backlash, and outright absurdity. People who get their animals certified as ESAs are often confused as to what the law actually states, so they will try to take their animal to, say, a grocery store. Legally, the only nonhuman animals allowed in such places are service dogs trained for a specific task, such as guiding a visually impaired person. Certifying an animal to become an ESA is far easier than certifying a service dog. (Dogs are the only animals recognized in the United States as capable of becoming genuine service animals.) A letter from a mental health-care provider specifying that the animal is integral to the patient's well-being is all that is needed. Despite what many people assume, a special vest or harness is not required. Nevertheless, companies such as the National Service Animal Registry,[6] with names that make them seem official and authentic, will gladly sell you "ESA kits," ranging from $50 to $500. The kits almost always include a vest, and you can also request a letter from a "healthcare provider." There are undoubtedly a great many fake ESAs out there, but the problem is that the Americans with Disabilities Act (ADA) prohibits any employee who encounters a person with a disability from asking that person to explain the disability. This includes persons who have ESAs. So it is difficult to estimate how many people truly need the animals for support and how many are gaming the system to get their dogs to fly with them for free. Confusion and potential corruption have resulted in the federal government cracking down on ESAs, and it has even been suggested that airlines might begin banning them altogether (Santuzzi et al. 2014; Brandt 2018).

Problems surrounding ESAs are not limited to legality and authenticity. Even more problematic, I think, are the underlying assumptions and ideologies present in discourse surrounding ESAs and the ever-changing legislation pertaining to them. The stricter the laws become and the more exclusions are put in place, the louder the pro-ESA movement shouts in retort. The move to destigmatize mental illness and to normalize treating it ought to be seen as a worthy cause, and it seems that ESAs are a good step in that direction. It says a lot about American greed and capitalism when people who suffer from mental illness, but who, because of lack of universal health care, are unable to afford treatment and then seek a way to get their beloved animal companion "certified" to travel with them, are scammed by a company promising to provide them with an "official" note from a "doctor." The backlash against ESAs by those who think the entire movement is a bunch of fraudulent people trying to save money is also illuminating, as it shows just how far we are from a world that takes seriously the increasing numbers of people affected by mental illness and how animals truly can be a part of that treatment process. Nevertheless, a pig who defecates on a flight and is a general nuisance to all on board cannot be tolerated. More important, there are genuine reasons why we might not want animals to be in public spaces. People can be highly allergic to dogs, for instance, and ironically, the very same dog that might help someone with PTSD could also trigger another person with the same disorder if, for example, that person suffered a vicious dog attack in the past. And finally, for all the compelling research about how companion animals like dogs think and feel, nowhere in any of these debates over ESAs do we find welfare considerations for those animals. Is it really in their best interest to fly with their human companions? In turning a blind eye to all these questions, humans subsequently overlook how neoliberal, capitalist, ableist, individualist, and species-chauvinist ideologies are bundled tightly together and frame our understanding of the issues.

Staunch advocates for and avid protestors against ESAs are often equally myopic in this way.

If instead we "stay with the trouble," as Haraway suggests, we are forced to confront issues such as these in their entire complexity. Part of doing this, I have argued, involves reframing how we think about dogs and our tight coupling with them. Engaging in this paradigm shift, both in philosophy and in cognitive ethology, not only allows for better theory but for better practice as well. By *rethinking thinking*—by recognizing that all sorts of collaborative dyads think together, play together, and make together—we have the possibility of an enactive and interspecies cognitive science. This new way of understanding cognition demands that we pay attention to all sorts of pairings, including human-human dyadic thinking, and that we not shy away from what such a radically different cognitive science might imply ethically, legally, or politically. Carefully *Minding Dogs* means more carefully *minding ourselves*.

NOTES

Introduction

1. Smith posted the 1993 footage, under the title "Drowning Puppy," on YouTube at https://www.youtube.com/watch?v=_PR5moWpa5Q.

2. See, respectively, Smith 1993; Sewall 2015; Sieczkowski 2013.

3. Though I do not endeavor to broach all these subjects in this book, my overarching thesis might help us think through issues arising as dogs take even more central roles in the lives of humans, including certified emotional support and/or therapy jobs. Likewise, our commitment to taking care of dogs is often complicated by bioethical considerations, as in the case of prescribing psychoactive drugs to pets, which can be seen as a means to force them to comply with our way of life. Understanding more fully how dogs think, as well as how we think about them, might shed light on how to navigate these complex issues.

4. It should be noted that throughout the book, "dogs" primarily refers to those members of *Canis familiaris* who live in extremely close proximity to humans—in their homes and backyards, and in constant contact with one or more humans we might call their family. Other dogs who live more liminally (see Donaldson and Kymlicka 2011 for more discussion about liminal species), such as the street dogs of India, most likely have quite different umwelten, so a lot of what I say about dog-human interfacing might not apply to all dogs in all contexts. Even street dogs have a lot of contact with humans, but I do not intend to generalize my arguments to dogs for whom the studies I discuss have not been replicated. It remains an open question, and one worth pursuing, to what extent "half-domesticated" dogs can think alongside humans in the same ways as their domesticated cousins.

5. Here I am referring to the way some dogs can perceive changes in blood sugar before diabetics even recognize this is happening, or the ways in which dogs trained

to assist with persons suffering from PTSD might detect anxiety or stress before the person becomes overtly aware of it.

6. See Hare and Woods 2013 and the website dognition.com for the at-home research project his lab has developed.

Chapter 1. Canine Minds

1. See Brosnan and Waal 2003; Marzluff and Angell 2005, 2012; Balcombe 2016; Godfrey-Smith 2016.

2. A closed Facebook group started by a philosopher, "Dogs who live with philosophers" is one of many internet destinations for dog lovers, though as with most of these sites, the content is much more fluff than serious philosophy.

3. For an overview, see Burkhardt 2005.

4. See Gallagher 2005; Varela et al. 1991. The definition of cognitive science, from Thagard (2018), is consistent with other definitions such as Gallagher and Zahavi's (2007, 1) and Andy Clark's (2013, 18–19).

5. See Nagel's 1974 discussion.

6. For more on the specifics of canine olfaction, see Craven et al. 2010.

7. From a radical behaviorist standpoint (see Skinner 1974), however, cognition itself *just is* behavior, so it might be argued that cognitive ethology adds nothing to the behaviorist paradigm. I am remaining relatively neutral in regard to this debate, as I see it to be a nominal issue. We are interested in explaining how dogs (and humans, and humans with dogs) think, and if that turns out to be really just an explanation of behaviors, so be it.

8. Edney's finding is perhaps most specious among those listed herein. See Strong et al. 1999 for critical commentary.

9. Guo et al. 2009; Smith et al. 2016. Significantly, Guo et al. find that dogs track only human faces, not any other type of face.

10. How this works, precisely, is a matter of much philosophical and psychological debate, and I speak about it more in the coming chapters. But it is fairly well agreed that humans are, from birth, quite adept at mimicking others' behaviors and then begin inferring mentality from those behaviors at very early ages.

11. However, it was most likely Thomas Reid who first introduced the problem of other minds into the discourse surrounding the issue. See Somerville 1989.

12. See also Overgaard 2006 for a discussion of how Wittgenstein approached this issue similarly.

13. See Dennett 2003c, 2007. It is worth noting that Dennett emphatically disagrees with Nagel's argument in "What Is It Like to Be a Bat?" but I think this is more a case of talking past one another, and I see what they offer as similar, at least in methodological terms.

Chapter 2. Thinking-with Dogs and Dismantling Standard Cognitive Science

1. See Waal 1996; Bekoff 2008; Bekoff and Pierce 2009; King 2013.
2. See Hare and Woods 2013 for a comprhensive review.
3. See Family Dog Project at http://familydogproject.elte.hu/. See also Miklósi et al. 2004.
4. Dolphins, especially of late, have been demonstrating unique abilities with regard to language. Though exactly *what* they are doing—merely making associations or actually learning the meanings of symbols—is a matter of contention, they have been shown to recognize symbols as naming objects, as actions to be performed, and even as names for themselves. See Hamilton 2011.
5. See Fouts et al. 1989 and Pollick et al. 2007 for an argument about evolutionary continuity between ape gestures and human language.
6. Chaser was trained for nearly five hours a day. Also, the question of whether she actually mapped words onto objects or was just responding to vocal patterns and extensive training is one that needs to be addressed as well. Pilley and Reid make a strong case that she was actually grasping meaning. When they introduced a novel toy to her "flock" and assigned a name to it, she was able to infer which toy they were asking for, despite having never seen the toy or heard its name. It is arguable that Chaser was doing something similar to what children do as they learn language, though children don't require the extensive and explicit training Chaser did.
7. As we saw in chapter 1, horses have recently been shown to exhibit this ability, but as with dogs, the results are not easily replicated, so there is some skepticism as to whether this is a universal capacity among each species. However, Proops et al. 2018 offers more compelling reasons to believe horses are quite attuned to humans' emotions.
8. For discussion of emotions as feelings, see the original work of William James (1884), from which most neo-Jamesian accounts take their lead (cf. Goldie 2000, 2004, 2009, 2010). For discussion of intentional theory and of emotions as perceptions, see Searle 1983; Kraut 1987; Prinz 2004. For cognitivist treatments, see Solomon 1973; Nussbaum 2001.
9. See Ratcliffe 2010; also Fonagy and Target 2007. For an excellent survey of background emotions and emotional regulation, see Varga and Krueger 2013.
10. See Varela et al. 1991 for an account of embodied cognition and enactivist stance. For a nice discussion of the extended mind hypothesis and its detractors, see Richard Menary's 2010 edited volume *The Extended Mind*.
11. Quoted in Cooper 2015. See also Pilley and Reid 2011.
12. Another point on which I disagree with Pilley is in referring to dogs as "pets." As the reader will note, I have taken care to resist language such as "pet/owner" that can be construed as treating the dog as an object to be owned. Likewise, if my argument is correct—namely, that dogs and humans are often co-constitutors of thinking, affect, and meaning—then it makes better sense to talk about dogs as companions, not pets.
13. Shipman and her sympathizers also claim that the very reason Neanderthals

did not survive but *Homo sapiens* did, is the supreme hunting skills of the latter. And it wasn't about tool use, unless we want to go so far as to call dogs "tools." Neanderthals made tools and weapons. What they failed to do was become intertwined with canids.

14. Some, though, will argue this; see Noë 2009. It depends on the particular flavor of enactivism being argued. As I have stated, there are very nuanced debates even among those claiming to be enactivists, but this need not concern us here.

15. There are many such stories, but one that is particularly compelling is that of Hachiko, an Akita in Japan. Hachiko always followed his primary guardian, Mr. Ueno, to the Shibuya train station from which he left for work each day, and then showed up around the time the train should return, to greet Ueno. One day Mr. Ueno died while at work, and so he never returned that afternoon. Hachiko continued to show up at the train station every day at the same time and wait. He did this until he died. The story is corroborated by several independent sources—see, for example, Newman 2004 and Turner 2004—and there is a statue erected at the station in his honor. Another more recent tale of canine grief (Hibbard 2011) is the case of fallen Navy SEAL John Tumilson, whose dog, Hawkeye, accompanied mourners at Tumilson's funeral. Hawkeye followed the casket down the aisle and then lay down under it for the entire service. Onlookers described the dog as full of sorrow, depressed, and in a state of grief. Although we cannot know for sure if Hawkeye understood his companion was in the casket, or if he even comprehended that Tumilson had died, we don't need a scientific experiment to prove that Hawkeye was grieving.

Chapter 3. Canine Mindreading and Interspecies Social Cognition

1. I refer to attachment here in the sense of "attachment theory" in psychology, as this paradigm has been utilized to explain infant and child behavior in various contexts as it emerges from secure and insecure relationships to others. We might say that attachment is prior to and a necessary condition for more complex social emotions such as love. In this way, dogs are capable of at least the former, according to research like Berns's, and likely capable of the latter as well.

2. The two terms are often used interchangeably among philosophers and psychologists, which can be confusing. It is important to keep in mind that "theory of mind" does not refer to an actual theory, but is rather a capacity to attribute mental states to others. I prefer "mindreading" to denote the skills deployed when we try to understand one another, and will generally use that term, which I think captures more of the general sense, while avoiding the overly linguistic connotations of "theory of mind." However, when rehearsing others' arguments, if they use the ToM nomenclature, I will also, just to preserve the general thrust of their argument. My point in doing this is purposeful, as will be evident as my argument unfolds.

3. Neuro-atypical children, such as those with autism, will often be delayed in passing this test, and in severe cases might never pass it.

4. Walker 1982. Flom et al. (2009) have also found that infants as young as six months can detect differences in canine vocalization, such as angry versus playful, and by twenty-four months they are able to properly match those vocalizations to dogs' facial expressions.

5. See Hutto and Myin 2012 for an overview and admittedly biased account, as their view is one of the most radically anti-representational among even the embodied/enacted/extended camp of philosophy of cognitive science.

6. Green et al. 2015 and Koehne et al. 2016 have both shown that interpersonal synchronous interaction increases empathetic responses in persons with ASD.

Chapter 4. Thinking-in-Playing

1. There are undoubtedly many other ways in which humans play that are not covered by this list. Playing a musical instrument is perhaps a type of play (indeed, we do not call it "working" the instrument), and yet this sort of play is often *hard work*. Lots of sports and games are similar, where the line between fun/play and hard/work becomes blurry. It is worth considering these points more in light of what we know about the role of improvisation in many forms of art and sport and how this free-form *playing-with* or bending the rules often results in creative new strategies or moves, or advancements in the game, and so forth. This idea will be discussed more as the analysis of human-dog play unfolds. I am thankful to one of the reviewers of the manuscript for bringing this point to my attention.

2. See Fagen 1981 for a review of mammalian versus nonmammalian play.

3. It is interesting to note that even the term "toy" can be considered anthropocentric. Since what it means to be a "plaything" is highly human-focused, assuming the object my dogs are "playing with" is a "toy" could be problematic. Perhaps to them the object is more like prey. Much the way my cat often appears to be "playing" with a half-dead chipmunk outside, the motive behind the action could be quite different, and calling the chipmunk a toy seems incorrect. (Plus, the cat's case would likely fail to meet Burghardt's requirements fully, given that this behavior could well be a survival-based stereotypy.) I take it that once we analyze most forms of supposed dog play, it will be clear that these do in fact count as genuinely playful activity, and hence calling the rope a "toy" is not a case of egregious anthropomorphism.

4. For more reason to believe what Darwin and Tesla are doing counts as play, see Bekoff 1995; Bauer and Smuts 2007.

5. For an overview of canid-canid play, however, see Bekoff and Byers 1998; Palagi et al. 2015; Byosiere et al. 2016.

6. See Mitchell 2001, 2004, and Mitchell et al. 2018 for the ways gender and familiarity impact the extent to which meanings are generated in play between humans and dogs.

Chapter 5. Dynamic Duos

1. Which is, incidentally, not entirely dissimilar from what bluefin tuna do in order to maximize thrust in the water, despite being relatively slow swimmers (see Triantafyllou and Triantafyllou 1995).

2. Though they are expected to "dance," they are not expected to dance in the style of, say, a ballerina or a Cunningham-trained modern dancer. It is an interesting, though much too tangential, question what makes a movement count as a "dance move" as opposed to something else, but as is the case with much of CI, many workshops or classes in which improvisation is used (on dancers or any other types of artists), some sort of expertise is assumed prior to the improvising, so the moves always already show up as dance moves. See Montero 2016 for a discussion of expertise and its supposed effects on the ability to move "without thought" and the phenomenon of "choking" under pressure.

3. Much of this of course has its roots in James J. Gibson, whose theory of affordances, as discussed in chapter 2, is central to many radically embodied or enactivist frameworks.

4. See A. Clark 2016, 85, for more discussion.

5. See Bruineberg et al. 2016 for a discussion.

6. Cf. Adams and Aizawa 2001, 2010. For a convincing reply to this objection specifically, see Menary 2006.

7. See, for example, the discussion in A. Clark 2016, 210.

8. For more discussion of examples such as these in human-dog working dyads, see Warren 2013.

Postscript

1. The championship game held yearly in American professional football.

2. The original "WeatherTech Super Bowl ad featuring Scout," which aired February 2, 2020, is on YouTube at https://www.youtube.com/watch?v=Fi2WwRJDiio.

3. It is worth noting that when AIBO is placed in the home of a living dog for an extended period, there might be reason to think that dogs begin to cynomorphize. One dog, when the robot was close to the dog's favorite toy, went over and snarled at AIBO, suggesting that the dog at least understood AIBO to be a potential threat.

4. The "uncanny valley" refers to the way humans have a threshold for how much anthropomorphism they can tolerate in a robot. In other words, as robots take on more humanlike traits, humans are more likely to treat them like humans, but if they become so humanlike that they begin to be indistinguishable from humans, then people tend to become uncomfortable.

5. See Grimm 2014 for more discussion.

6. The National Service Dog Registry can be found at https://www.nsarco.com/esa-registration-kit-options.html.

WORKS CITED

Adams, Fred, and Ken Aizawa. 2001. "The Bounds of Cognition." *Philosophical Psychology* 14 (1): 43–64.
———. 2010. "Defending the Bounds of Cognition." In Menary 2010, 67–80.
Adamson, Lauren B., and Janet E. Frick. 2003. "The Still Face: A History of a Shared Experimental Paradigm." *Infancy* 4 (4): 451–73.
Agamben, Giorgio. 2004. *The Open: Man and Animal*. Stanford University Press.
Albert, David J. 2011. "What's on the Mind of a Jellyfish? A Review of Behavioural Observations on *Aurelia sp.* Jellyfish." *Neuroscience and Biobehavioral Reviews* 35 (3): 474–82.
Allen, Colin, and Marc Bekoff. 1997. *Species of Mind: The Philosophy and Biology of Cognitive Ethology*. MIT Press.
Andrews, Kristin. 2005. "Chimpanzee Theory of Mind: Looking in All the Wrong Places?" *Mind and Language* 20 (5): 521–36.
———. 2011. "Beyond Anthropomorphism: Attributing Psychological Properties to Animals." In *The Oxford Handbook of Animal Ethics*, edited by Tom L. Beauchamp and R. G. Frey, 469–94. Oxford University Press.
———. 2012. *Do Apes Read Minds?: Toward a New Folk Psychology*. MIT Press.
———. 2014. *The Animal Mind: An Introduction to the Philosophy of Animal Cognition*. Routledge.
Andrews, Kristin, and Brian Huss. 2014. "Anthropomorphism, Anthropectomy, and the Null Hypothesis." *Biology and Philosophy* 29 (5): 711–29.
APPA [American Pet Products Association]. 2018. *2017–2018 APPA National Pet Owners Survey Debut*. APPA.
Arbib, Michael A. 2002. "The Mirror System, Imitation, and the Evolution of Language." In *Imitation in Animals and Artifacts*, edited by Kerstin Dautenhahn and Chrystopher L. Nehaniv, 229–80. MIT Press.

Aristotle. 1946. *The Politics of Aristotle*. Edited and translated by Ernest Baker. Clarendon Press.

Arnott, S. R., L. Thaler, J. L. Milne, D. Kish, and M. A. Goodale. 2013. "Shape-Specific Activation of Occipital Cortex in an Early Blind Echolocation Expert." *Neuropsychologia* 51 (5): 938–49.

Aust, Ulrike, Friederike Range, Michael Steurer, and Ludwig Huber. 2008. "Inferential Reasoning by Exclusion in Pigeons, Dogs, and Humans." *Animal Cognition* 11 (4): 587–97.

Ayre, David J., and Richard K. Grosberg. 1995. "Aggression, Habituation, and Clonal Coexistence in the Sea Anemone *Anthopleura elegantissima*." *American Naturalist* 146 (3): 427–53.

———. 1996. "Effects of Social Organization on Inter-Clonal Dominance Relationships in the Sea Anemone *Anthopleura elegantissima*." *Animal Behaviour* 51 (6): 1233–45.

Balcombe, Jonathan. 2016. *What a Fish Knows: The Inner Lives of Our Underwater Cousins*. Farrar, Straus and Giroux.

Baron-Cohen, Simon. 1997. *Mindblindness: An Essay on Autism and Theory of Mind*. Rev. ed. MIT Press.

———. 2006. "Mindreading: Evidence for Both Innate and Acquired Factors." *Anthropological Psychology* 17: 26–27.

Baron-Cohen, Simon, Alan M. Leslie, and Uta Frith. 1985. "Does the Autistic Child Have a 'Theory of Mind'?" *Cognition* 21 (1): 37–46.

Barrett, L., P. Henzi, and D. Rendall. 2007. "Social Brains, Simple Minds: Does Social Complexity Really Require Cognitive Complexity?" *Philosophical Transactions of the Royal Society of London B: Biological Sciences* 362 (1480): 561–575.

Barrett, Lisa Feldman, and W. Kyle Simmons. 2015. "Interoceptive Predictions in the Brain." *Nature Reviews Neuroscience* 16 (7): 419–29.

Bhattacharjee, Debottam, Nikhil Dev N, Shreya Gupta, Shubhra Sau, Rohan Sarkar, Arpita Biswas, Arunita Banerjee, Daisy Babu, Diksha Mehta, and Anindita Bhadra. 2017. "Free-Ranging Dogs Show Age Related Plasticity in Their Ability to Follow Human Pointing." *PLoS One* 12 (7): e0180643.

Bauer, Erika B., and Barbara B. Smuts. 2007. "Cooperation and Competition during Dyadic Play in Domestic Dogs, *Canis familiaris*." *Animal Behaviour* 73 (3): 489–99.

BBC News. 2018. "What's the Kindest Way to Kill a Lobster?" January 11. https://www.bbc.com/news/world-europe-42647341.

Bekoff, Marc. 1984. "Social Play Behavior." *Bioscience* 34 (4): 228–33.

———. 1995. "Play Signals as Punctuation: The Structure of Social Play in Canids." *Behaviour* 132 (5/6): 419–29.

———. 2008. *The Emotional Lives of Animals: A Leading Scientist Explores Animal Joy, Sorrow, and Empathy—and Why They Matter*. New World Library.

Bekoff, Marc, and Colin Allen. 1997. "Cognitive Ethology: Slayers, Skeptics, and Proponents." In *Anthropomorphism, Anecdotes, and Animals*, edited by Robert W. Mitchell, Nicholas S. Thompson, and H. Lyn Miles, 313–34. SUNY Press.

Bekoff, Marc, Colin Allen, and Gordon M. Burghardt, eds. 2002. *The Cognitive Animal: Empirical and Theoretical Perspectives on Animal Cognition*. MIT Press.

Bekoff, Marc, and John A. Byers. 1998. *Animal Play: Evolutionary, Comparative, and Ecological Perspectives*. Cambridge University Press.

Bekoff, Marc, and Jessica Pierce. 2009. *Wild Justice: The Moral Lives of Animals*. University of Chicago Press.

Bergelson, Elika, and Daniel Swingley. 2012. "At 6–9 Months, Human Infants Know the Meanings of Many Common Nouns." *Proceedings of the National Academy of Sciences of the United States of America* 109 (9): 3253–58.

Berk, Laura E., Trisha D. Mann, and Amy T. Ogan. 2006. "Make-Believe Play: Wellspring for Development of Self-Regulation." In *Play = Learning: How Play Motivates and Enhances Children's Cognitive and Social-Emotional Growth*, edited by Dorothy G. Singer, Roberta Michnick Golinkoff, and Kathy Hirsh-Pasek, 74–100. Oxford University Press.

Bermúdez, José Luis. 2003. "Two Approaches to the Nature of Thought." In *Thinking without Words*, 13–33. Oxford University Press.

Berns, Gregory. 2013. *How Dogs Love Us: A Neuroscientist and His Adopted Dog Decode the Canine Brain*. Houghton Mifflin Harcourt.

Berns, Gregory S., Andrew M. Brooks, and Mark Spivak. 2015. "Scent of the Familiar: An fMRI Study of Canine Brain Responses to Familiar and Unfamiliar Human and Dog Odors." *Behavioural Processes* 110 (January): 37–46.

Berr, Jonathan. 2015. "The High Costs—and Heartbreak—of Pet Cancer." *CBS News*, December 18. https://www.cbsnews.com/news/the-heartbreak-and-high-costs-of-pet-cancer/.

Bordenstein Seth R., and Kevin R. Theis. 2015. "Host Biology in Light of the Microbiome: Ten Principles of Holobionts and Hologenomes." *PLoS Biol* 13 (8): e1002226. doi:10.1371/journal.pbio.1002226.

Boyd, Jacqueline. 2016. "How Did Moscow's Stray Dogs Learn to Navigate the Metro?" *The Conversation*, February 18. https://theconversation.com/how-did-moscows-stray-dogs-learn-to-navigate-the-metro-54790.

Braitman, Laurel. 2014. *Animal Madness: How Anxious Dogs, Compulsive Parrots, and Elephants in Recovery Help Us Understand Ourselves*. Simon and Schuster.

Brandt, Sydney. 2018. "Airlines Crack Down on Emotional Support Animals in Plane Cabins." ABC News, July 25. https://abcnews.go.com/US/airlines-crack-emotional-support-animals-plane-cabins/story?id=56791124.

Brauer, Juliane, Juliane Kaminski, Julia Riedel, Josep Call, and Michael Tomasello. 2006. "Making Inferences about the Location of Hidden Food: Social Dog, Causal Ape." *Journal of Comparative Psychology* 120 (1): 38–47.

Broadhead, Pat, Justine Howard, and Elizabeth Wood, eds. 2010. *Play and Learning in the Early Years: From Research to Practice*. SAGE.

Brooker, Rebecca J., Kristin A. Buss, Kathryn Lemery-Chalfant, Nazan Aksan, Richard J. Davidson, and H. Hill Goldsmith. 2013. "The Development of Stranger Fear in Infancy and Toddlerhood: Normative Development, Individual Differences, Antecedents, and Outcomes." *Developmental Science* 16 (6): 864–78.

Brooks, Rodney A. 1999. *Cambrian Intelligence: The Early History of the New AI*. MIT Press.

Brosnan, Sarah F., and Frans B. M. de Waal. 2003. "Monkeys Reject Unequal Pay." *Nature* 425 (6955): 297–99.

Bruineberg, Jelle, Julian Kiverstein, and Erik Rietveld. 2016. "The Anticipating Brain Is Not a Scientist: The Free-Energy Principle from an Ecological-Enactive Perspective." *Synthese*, October. doi:10.1007/s11229-016-1239-1.

Bugnyar, Thomas, and Bernd Heinrich. 2005. "Ravens, *Corvus corax*, Differentiate between Knowledgeable and Ignorant Competitors." *Proceedings of the Royal Society B: Biological Sciences* 272 (1573): 1641–46.

Burghardt, Gordon M. 1991. "Cognitive Ethology and Critical Anthropomorphism: A Snake with Two Heads and Hog-Nose Snakes That Play Dead." In *Cognitive Ethology: The Minds of Other Animals: Essays in Honor of Donald R. Griffin*, edited by Carolyn R. Ristau, 73–110. Psychology Press.

———. 2005. *The Genesis of Animal Play: Testing the Limits*. MIT Press.

———. 2007. "Critical Anthropomorphism, Uncritical Anthropocentrism, and Naïve Nominalism." *Comparative Cognition and Behavior Reviews* 2: 136–38. http://www.jennifervonk.com/uploads/7/7/3/2/7732985/burghardt_2007.pdf.

Burghardt, Gordon M., Brian Ward, and Roger Rosscoe. 1996. "Problem of Reptile Play: Environmental Enrichment and Play Behavior in a Captive Nile Soft-Shelled Turtle, *Trionyx triunguis*." *Zoo Biology* 15 (3): 223–38. doi:10.1002/(SICI)1098-2361(1996)15:3<223::AID-ZOO3>3.0.CO;2-D.

Burkhardt, Richard W., Jr. 2005. *Patterns of Behavior: Konrad Lorenz, Niko Tinbergen, and the Founding of Ethology*. University of Chicago Press.

Butterfill, Stephen A., and Ian A. Apperly. 2013. "How to Construct a Minimal Theory of Mind." *Mind and Language* 28 (5): 606–37.

Buytendijk, F. J. J. 1936. *The Mind of the Dog*. Translated by Lilian A. Clare. Houghton Mifflin.

Byosiere, Sarah-Elizabeth, Julia Espinosa, and Barbara Smuts. 2016. "Investigating the Function of Play Bows in Adult Pet Dogs (*Canis lupus familiaris*)." *Behavioural Processes* 125 (April): 106–13.

Calvo-Merino, B., D. E. Glaser, J. Grèzes, R. E. Passingham, and P. Haggard. 2005. "Action Observation and Acquired Motor Skills: An FMRI Study with Expert Dancers." *Cerebral Cortex* 15 (8): 1243–49.

Calvo-Merino, Beatriz, Julie Grèzes, Daniel E. Glaser, Richard E. Passingham, and Patrick Haggard. 2006. "Seeing or Doing? Influence of Visual and Motor Familiarity in Action Observation." *Current Biology* 16 (19): 1905–10.

Campos, Joseph J., Susan Hiatt, Douglas Ramsay, Charlotte Henderson, and Marilyn Svejda. 1978. "The Emergence of Fear on the Visual Cliff." In *The Development of Affect*, edited by Michael Lewis and Leonard A. Rosenblum, 149–82. Springer / Plenum.

Carruthers, Peter. 2013. "Mindreading in Infancy." *Mind and Language* 28 (2): 141–72.

Chemero, Anthony. 2006. "Information and Direct Perception: A New Approach." In *Advanced Issues in Cognitive Science and Semiotics*, edited by Priscila Farias and João Queiroz, 59–72. Shaker Verlag Aachen.

Cheng, Ken. 1995. "Landmark-Based Spatial Memory in the Pigeon." In *The Psychology of Learning and Motivation*, edited by Douglas L. Medin, 33:1–21. Academic Press.

———. 2018. "Cognition Beyond Representation: Varieties of Situated Cognition in Animals." *Comparative Cognition and Behavior Reviews* 13: 1–20.

Chevalier-Skolnikoff, Suzanne. 1986. "An Exploration of the Ontogeny of Deception in Human Beings and Nonhuman Primates." In Mitchell and Thompson 1986a, 205–20.

Chomsky, Noam. 2006. *Language and Mind*. 3rd ed. Cambridge University Press.

Churchland, Paul M. 1992. "Activation Vectors versus Propositional Attitudes: How the Brain Represents Reality." *Philosophy and Phenomenological Research* 52 (2): 419–24.

Clark, Andy. 1993. *Associative Engines: Connectionism, Concepts, and Representational Change*. MIT Press.

———. 1998. *Being There: Putting Brain, Body, and World Together Again*. MIT Press.

———. 2003. *Natural-Born Cyborgs: Minds, Technologies, and the Future of Human Intelligence*. Oxford University Press.

———. 2008. *Supersizing the Mind: Embodiment, Action, and Cognitive Extension*. Oxford University Press.

———. 2013. *Mindware: An Introduction to the Philosophy of Cognitive Science*. 2nd ed. Oxford University Press.

———. 2016. *Surfing Uncertainty: Prediction, Action, and the Embodied Mind*. Oxford University Press.

Clark, Andy, and David J. Chalmers. 1998. "The Extended Mind." *Analysis* 58 (1): 7–19.

Clark, Cindy Dell. 2007. "Therapeutic Advantages of Play." In *Play and Development: Evolutionary, Sociocultural, and Functional Perspectives*, edited by Artin Göncü and Suzanne Gaskins, 275–93. Lawrence Erlbaum.

Clayton, Nicola S., Joanna M. Dally, and Nathan J. Emery. 2007. "Social Cognition by Food-Caching Corvids: The Western Scrub-Jay as a Natural Psychologist." *Philosophical Transactions of the Royal Society of London B: Biological Sciences* 362 (1480): 507–22.

Coelho, Luis Pedro, Jens Roat Kultima, Paul Igor Costea, et al. 2018. "Similarity of the Dog and Human Gut Microbiomes in Gene Content and Response to Diet." *Microbiome* 6: 72.

Coelho, Saroja. 2013. "Dolphins Gain Unprecedented Protection in India." Deutsche Welle. May 24. https://www.dw.com/en/dolphins-gain-unprecedented-protection-in-india/a-16834519.

Cohen, Andrew J., Hartley Brody, German Patino, et al. 2019. "Use of an Online Crowdfunding Platform for Unmet Financial Obligations in Cancer Care." *JAMA Internal Medicine* 179 (12): 1717–20. doi:10.1001/jamainternmed.2019.3330.

Cooper, Anderson. 2015. "The Smartest Dog in the World." *60 Minutes*, CBS News, June 14. https://www.cbsnews.com/news/smart-dog-anderson-cooper-60-minutes/.

Coppinger, Raymond, and Lorna Coppinger. 2001. *Dogs: A Startling New Understanding of Canine Origin, Behavior, and Evolution*. Scribner.

Cornu, Jean-Nicolas, Géraldine Cancel-Tassin, Valérie Ondet, Caroline Girardet, and Olivier Cussenot. 2011. "Olfactory Detection of Prostate Cancer by Dogs Sniffing Urine: A Step Forward in Early Diagnosis." *European Urology* 59 (2): 197–201.

Cort, Julia. 2011. "How Smart Are Dogs?" Segment of *How Smart Are Animals?* Directed by Alexis Bloom, Julia Cort, Rushmore DeNooyer, Irene Pepperberg, Joshua Seftel, and Dana Rae Warren. *Nova Science Now*, February 9.

Craven, Brent A., Eric G. Paterson, and Gary S. Settles. 2010. "The Fluid Dynamics of Canine Olfaction: Unique Nasal Airflow Patterns as an Explanation of Macrosmia." *Journal of the Royal Society Interface* 7 (47): 933–43.

Cronin, Helena. 1992. "What Do Animals Want?" Review of *Animal Minds*, by Donald J. Griffin (University of Chicago Press, 1992). *New York Times Book Review*, November 1.

Currie, Gregory. 2007. "Narrative Frameworks." In *Narrative and Understanding Persons*, edited by Daniel D. Hutto, 17–42. Royal Institute of Philosophy Supplements 60. Cambridge University Press.

Dally, Joanna M., Nathan J. Emery, and Nicola S. Clayton. 2004. "Cache Protection Strategies by Western Scrub-Jays (*Aphelocoma californica*): Hiding Food in the Shade." *Proceedings of the Royal Society B: Biological Sciences* 271 Suppl 6 (December): S387–90.

Daly, Anya. 2014. "Primary Intersubjectivity: Empathy, Affective Reversibility, 'Self-Affection' and the Primordial 'We.'" *Topoi* 33 (1): 227–41.

Darwin, Charles. 1859. *On the Origin of Species by Means of Natural Selection, or, The Preservation of Favoured Races in the Struggle for Life*. John Murray.

———. 1871. *The Descent of Man, and Selection in Relation to Sex*. John Murray.

Daston, Lorraine, and Gregg Mitman. 2005. *Thinking with Animals: New Perspectives on Anthropomorphism*. Columbia University Press.

De Dreu, Carsten K. W., Lindred L. Greer, Gerben A. van Kleef, Shaul Shalvi, and Michel J. J. Handgraaf. 2011. "Oxytocin Promotes Human Ethnocentrism." *Proceedings of the National Academy of Sciences of the United States of America* 108 (4): 1262–66.

De Jaegher, Hanne, and Ezequiel Di Paolo. 2007. "Participatory Sense-Making." *Phenomenology and the Cognitive Sciences* 6 (4): 485–507.

De Jaegher, Hanne, Ezequiel Di Paolo, and Shaun Gallagher. 2010. "Can Social Interaction Constitute Social Cognition?" *Trends in Cognitive Sciences* 14 (10): 441–47.

Dempster, M. Beth. 1998. "A Self-Organizing Systems Perspective on Planning for Sustainability." Master's thesis, University of Waterloo (Ontario).

Dennett, Daniel C. 1987. *The Intentional Stance*. MIT Press.

———. 1991. "Real Patterns." *Journal of Philosophy* 88 (1): 27–51.

———. 1996. *Kinds of Minds: Toward an Understanding of Consciousness*. Basic Books.

———. 2003a. "Explaining the 'Magic' of Consciousness." *Journal of Cultural and Evolutionary Psychology* 1 (1): 7–19.

———. 2003b. "The Illusion of Consciousness." TED talk. https://www.ted.com/talks/dan_dennett_the_illusion_of_consciousness/transcript?language=en#:~:text=If%20you're%20going%20to,I%20attempt%20to%20explain%20consciousness.

———. 2003c. "Who's on First? Heterophenomenology Explained." *Journal of Consciousness Studies* 10 (9): 19–30.
———. 2007. "Heterophenomenology Reconsidered." *Phenomenology and the Cognitive Sciences* 6 (1–2): 247–70.
Descartes, Réné. (1637) 1999. *Discourse on Method, and Related Writings*. Translated by Desmond M. Clarke. Penguin.
Despret, Vinciane. 2008. "The Becomings of Subjectivity in Animal Worlds." *Subjectivity* 23 (1): 123–39.
Dewey, John. 1884. "The New Psychology." *Andover Review* 2: 278–89.
———. 1938. *Logic: The Theory of Inquiry*. Henry Holt.
Di Paolo, Ezequiel, and Hanne De Jaegher. 2012. "The Interactive Brain Hypothesis." *Frontiers in Human Neuroscience* 6 (163): 1–16.
Di Paolo, Ezequiel, Marieke Rohde, and Hanne De Jaegher. 2010. "Horizons for the Enactive Mind: Values, Social Interaction, and Play." In *Enaction: Toward a New Paradigm for Cognitive Science*, edited by John Stewart, Olivier Gapenne, and Ezequiel A. Di Paolo, 33–87. MIT Press.
Donaldson, Sue, and Will Kymlicka. 2011. *Zoopolis: A Political Theory of Animal Rights*. Oxford University Press.
Dorey, Nicole R., Monique A. R. Udell, and Clive D. L. Wynne. 2009. "Breed Differences in Dogs Sensitivity to Human Points: A Meta-Analysis." *Behavioural Processes* 81 (3): 409–15.
———. 2010. "When Do Domestic Dogs, *Canis familiaris*, Start to Understand Human Pointing? The Role of Ontogeny in the Development of Interspecies Communication." *Animal Behaviour* 79 (1): 37–41.
Du, Shichuan, Yong Tao, and Aleix M. Martinez. 2014. "Compound Facial Expressions of Emotion." *Proceedings of the National Academy of Sciences of the United States of America* 111 (15): E1454–62.
Edney, A. T. 1993. "Dogs and Human Epilepsy." *Veterinary Record* 132 (14): 337–38.
Ehmann, R., E. Boedeker, U. Friedrich, J. Sagert, J. Dippon, G. Friedel, and T. Walles. 2012. "Canine Scent Detection in the Diagnosis of Lung Cancer: Revisiting a Puzzling Phenomenon." *European Respiratory Journal* 39 (3): 669–76.
Elster, Jon. 1984. *Ulysses and the Sirens: Studies in Rationality and Irrationality*. Rev. ed. Cambridge University Press.
———. 2000. *Ulysses Unbound: Studies in Rationality, Precommitment, and Constraints*. Cambridge University Press.
Erdőhegyi, Ágnes, József Topál, Zsófia Virányi, and Ádám Miklósi. 2007. "Dog-Logic: Inferential Reasoning in a Two-Way Choice Task and Its Restricted Use." *Animal Behaviour* 74 (4): 725–37.
Fagen, Robert. 1981. *Animal Play Behavior*. Oxford University Press.
Farroni, Teresa, Mark H. Johnson, Enrica Menon, Luisa Zulian, Dino Faraguna, and Gergely Csibra. 2005. "Newborns' Preference for Face-Relevant Stimuli: Effects of Contrast Polarity." *Proceedings of the National Academy of Sciences of the United States of America* 102 (47): 17245–50.

Fernyhough, Charles, and Emma Fradley. 2005. "Private Speech on an Executive Task: Relations with Task Difficulty and Task Performance." *Cognitive Development* 20 (1): 103–20.

Fewkes, J. Walter. 1880. "The Siphonophores. I. The Anatomy and Development of Agalma." *American Naturalist* 14 (9): 617–30. http://www.journals.uchicago.edu/doi/pdfplus/10.1086/272637.

Fisher, T. 2020. "Brain Scans Reveal What Dogs Really Think About Us." Mic, February 13. https://mic.com/articles/104474/brain-scans-reveal-what-dogs-really-think-of-us#.zS5Yy4TxZ.

Flom, Ross, Heather Whipple, and Daniel Hyde. 2009. "Infants' Intermodal Perception of Canine (*Canis familairis*) Facial Expressions and Vocalizations." *Developmental Psychology* 45 (4): 1143–51.

Flombaum, Jonathan I., and Laurie R. Santos. 2005. "Rhesus Monkeys Attribute Perceptions to Others." *Current Biology* 15 (5): 447–52.

Fodor, J. A. 1980. "Methodological Solipsism Considered as a Research Strategy in Cognitive Psychology." *Behavioral and Brain Sciences* 3 (1): 63–73.

Fonagy, Peter, and Mary Target. 2007. "Playing with Reality: IV. A Theory of External Reality Rooted in Intersubjectivity." *International Journal of Psychoanalysis* 88 (pt. 4): 917–37.

Fouts, Roger S., Deborah H. Fouts, and Thomas E. Van Cantfort. 1989. "The Infant Loulis Learns Signs from Cross-Fostered Chimpanzees." In *Teaching Sign Language to Chimpanzees*, edited by R. Allen Gardner, Beatrix T. Gardner, and Thomas E. Van Cantfort, 280–92. SUNY Press.

Friston, Karl J., and Christopher D. Frith. 2015. "Active Inference, Communication and Hermeneutics." *Cortex* 68 (July): 129–43.

Frith, Chris D., and Thomas Schwartz Wentzer. 2013. "Neural Hermeneutics." *Encyclopedia of Philosophy and the Social Sciences*, edited by Byron Kaldis, 2:657–59. SAGE.

Froling, Joan. 2001. Assistance Dog Tasks." International Association of Assistance Dog Partners. https://www.iaadp.org/tasks.html.

———. 2009. "Service Dog Tasks for Psychiatric Disabilities: Tasks to Mitigate Certain Disabling Illnesses Classified as Mental Impairments under The Americans with Disabilities Act." https://www.iaadp.org/psd_tasks.html.

Fromberg, Doris Pronin, and Doris Bergen, eds. 2006. *Play from Birth to Twelve: Contexts, Perspectives, and Meanings*. 2nd ed. Routledge.

Fuchs, Thomas. 2002. "The Phenomenology of Shame, Guilt and the Body in Body Dysmorphic Disorder and Depression." *Journal of Phenomenological Psychology* 33 (2): 223–43.

Gácsi, Márta, Edina Kara, Bea Belényi, József Topál, and Ádám Miklósi. 2009. "The Effect of Development and Individual Differences in Pointing Comprehension of Dogs." *Animal Cognition* 12 (3): 471–79.

Gadamer, Hans-Georg. 1989. *Truth and Method*. 2nd ed. Translation revised by Joel Weinsheimer and Donald G. Marshall. Continuum.

Gallagher, Shaun. 2001. "The Practice of Mind: Theory, Simulation or Primary Interaction?" *Journal of Consciousness Studies* 8 (5–7): 83–108.

———. 2005. *How the Body Shapes the Mind*. Clarendon Press.
———. 2008a. "Direct Perception in the Intersubjective Context." *Consciousness and Cognition* 17 (2): 535–43.
———. 2008b. "Inference or Interaction: Social Cognition without Precursors." *Philosophical Explorations* 11 (3): 163–74.
———. 2013. "The Socially Extended Mind." *Cognitive Systems Research* 25–26 (December): 4–12.
———. 2020. "What in the World: Conversation and Things in Context." In *Minimal Co-operation and Shared Agency*, edited by Anika Fiebich, 59–70. Springer.
Gallagher, Shaun, and Micah Allen. 2016. "Active Inference, Enactivism and the Hermeneutics of Social Cognition." *Synthese* 195: 2627–48 (2018). doi:10.1007/s11229-016-1269-8.
Gallagher, Shaun, and Daniel D. Hutto. 2008. "Understanding Others through Primary Interaction and Narrative Practice." In *The Shared Mind: Perspectives on Intersubjectivity*, edited by Jordan Zlatev, Timothy P. Racine, Chris Sinha, and Esa Itkonen, 17–38. Converging Evidence in Language and Communication Research 12. John Benjamins.
Gallagher, Shaun, and Andrew N. Meltzoff. 1996. "The Earliest Sense of Self and Others: Merleau-Ponty and Recent Developmental Studies." *Philosophical Psychology* 9 (2): 211–33. doi:10.1080/09515089608573181.
Gallagher, Shaun, and Tailer G. Ransom. 2016. "Artifacting Minds: Material Engagement Theory and Joint Action." In *Embodiment in Evolution and Culture*, edited by Gregor Etzelmüller and Christian Tewes, 337–51. Mohr Siebeck.
Gallagher, Shaun, and Dan Zahavi. 2007. *The Phenomenological Mind: An Introduction to Philosophy of Mind and Cognitive Science*. Routledge.
Gallese, Vittorio, and Alvin Goldman. 1998. "Mirror Neurons and the Simulation Theory of Mind-Reading." *Trends in Cognitive Sciences* 2 (12): 493–501.
Gallese, Vittorio, Christian Keysers, and Giacomo Rizzolatti. 2004. "A Unifying View of the Basis of Social Cognition." *Trends in Cognitive Sciences* 8 (9): 396–403.
Gamble, Jennifer R., and Daniel A. Cristol. 2002. "Drop-Catch Behaviour Is Play in Herring Gulls, *Larus argentatus*." *Animal Behaviour* 63 (2): 339–45.
Geertz, Clifford. 1973. "Thick Description: Toward an Interpretive Theory of Culture." In *The Interpretation of Cultures: Selected Essays*, 310–23. Basic Books.
Gibson, Eleanor J., and Richard D. Walk. 1960. "The 'Visual Cliff.'" *Scientific American* 202 (4): 64–71.
Gibson, James J. 1977. "The Theory of Affordances." In *Perceiving, Acting, and Knowing*, edited by Robert E. Shaw and John Bransford, 67–82. Lawrence Erlbaum.
———. 1979. *The Ecological Approach to Visual Perception*. Houghton Mifflin.
Gilbert, Margaret. 1990. "Walking Together: A Paradigmatic Social Phenomenon." *Midwest Studies in Philosophy* 15 (1): 1–14.
Godfrey-Smith, Peter. 2016. *Other Minds: The Octopus, the Sea, and the Deep Origins of Consciousness*. Farrar, Straus and Giroux.
Goldie, Peter. 2000. *The Emotions: A Philosophical Exploration*. Clarendon Press.
———. 2004. "Emotion, Feeling, and Knowledge of the World." In *Thinking about Feel-

ing: Contemporary Philosophers on Emotions, edited by Robert C. Solomon, 91–105. Oxford University Press.

———. 2009. "Getting Feelings into Emotional Experience in the Right Way." *Emotion Review* 1 (3): 232–39.

———, ed. 2010. *The Oxford Handbook of Philosophy of Emotion*. Oxford University Press.

Goldman, Alvin I. 2006. *Simulating Minds: The Philosophy, Psychology, and Neuroscience of Mindreading*. Oxford University Press.

———. 2009. "Mirroring, Simulating and Mindreading." *Mind and Language* 24 (2): 235–52.

Gopnik, Alison, and Janet W. Astington. 1988. "Children's Understanding of Representational Change and Its Relation to the Understanding of False Belief and the Appearance-Reality Distinction." *Child Development* 59 (1): 26–37.

Gopnik, Alison, and Andrew N. Meltzoff. 1996. *Words, Thoughts, and Theories*. MIT Press.

Gopnik, Alison, Andrew N. Meltzoff, and Patricia K. Kuhl. 1999. *The Scientist in the Crib: Minds, Brains, and How Children Learn*. William Morrow.

Gopnik, Alison, and Henry M. Wellman. 1992. "Why the Child's Theory of Mind Really Is a Theory." *Mind and Language* 7 (1–2): 145–71.

Gordon, Robert M. 1986. "Folk Psychology as Simulation." *Mind and Language* 1 (2): 158–71.

Green, Jonathan, Tony Charman, Andrew Pickles, Ming W. Wan, Mayada Elsabbagh, Vicky Slonims, Carol Taylor, et al. 2015. "Parent-Mediated Intervention versus No Intervention for Infants at High Risk of Autism: A Parallel, Single-Blind, Randomised Trial." *Lancet Psychiatry* 2 (2): 133–40.

Griffin, Donald R. 1976. *The Question of Animal Awareness: Evolutionary Continuity of Mental Experience*. Rockefeller University Press.

———. 1984. *Animal Thinking*. Harvard University Press.

Grimes, John. 1996. "On the Failure to Detect Changes in Scenes across Saccades." In *Perception*, edited by Kathleen Akins, 89–110. Vancouver Studies in Cognitive Science 5. Oxford University Press.

Grimm, David. 2014. *Citizen Canine: Our Evolving Relationship with Cats and Dogs*. PublicAffairs.

Guo, Kun, Kerstin Meints, Charlotte Hall, Sophie Hall, and Daniel Mills. 2009. "Left Gaze Bias in Humans, Rhesus Monkeys and Domestic Dogs." *Animal Cognition* 12 (3): 409–18.

Guzmán, Yomayra F., Natalie C. Tronson, Vladimir Jovasevic, Keisuke Sato, Anita L. Guedea, Hiroaki Mizukami, Katsuhiko Nishimori, and Jelena Radulovic. 2013. "Fear-Enhancing Effects of Septal Oxytocin Receptors." *Nature Neuroscience* 16 (9): 1185–87.

Haddock, Steven H. D., Casey W. Dunn, Philip R. Pugh, and Christine E. Schnitzler. 2005. "Bioluminescent and Red-Fluorescent Lures in a Deep-Sea Siphonophore." *Science* 309 (5732): 263.

Hamilton, Doug. 2011. "Dolphin Reading Test." *Nova*, February 8. https://www.pbs.org/wgbh/nova/video/dolphin-reading-test.

Haralson, J. V., C. I. Groff, and S. J. Haralson. 1975. "Classical Conditioning in the Sea Anemone, Cribrina Xanthogrammica." *Physiology and Behavior* 15 (5): 455–60.

Haraway, Donna. (1985) 2006. "A Cyborg Manifesto: Science, Technology, and Socialist-Feminism in the Late 20th Century." In *The International Handbook of Virtual Learning Environments*, edited by Joel Weiss, 117–58. Springer.

———. 1988. "Situated Knowledges: The Science Question in Feminism and the Privilege of Partial Perspective." *Feminist Studies* 14 (3): 575–99.

———. 2008. *When Species Meet*. University of Minnesota Press.

———. 2016. *Staying with the Trouble: Making Kin in the Chthulucene*. Duke University Press.

Hare, Brian, Michelle Brown, Christina Williamson, and Michael Tomasello. 2002. "The Domestication of Social Cognition in Dogs." *Science* 298 (5598): 1634–36.

Hare, Brian, Josep Call, and Michael Tomasello. 1998. "Communication of Food Location between Human and Dog (*Canis familiaris*)." *Evolution of Communication* 2 (1): 137–59.

———. 2001. "Do Chimpanzees Know What Conspecifics Know?" *Animal Behaviour* 61 (1): 139–51.

Hare, Brian, Alexandra Rosati, Juliane Kaminski, Juliane Brauer, Josep Call, and Michael Tomasello. 2010. "The Domestication Hypothesis for Dogs' Skills with Human Communication: A Response to Udell et al. (2008) and Wynne et al. (2008)." *Animal Behaviour* 79 (2): e1–e6.

Hare, Brian, and Michael Tomasello. 1999. "Domestic Dogs (*Canis familiaris*) Use Human and Conspecific Social Cues to Locate Hidden Food." *Journal of Comparative Psychology* 113 (2): 173–77.

Hare, Brian, Victoria Wobber, and Richard Wrangham. 2012. "The Self-Domestication Hypothesis: Evolution of Bonobo Psychology Is Due to Selection against Aggression." *Animal Behaviour* 83 (3): 573–85.

Hare, Brian, and Vanessa Woods. 2013. *The Genius of Dogs: How Dogs Are Smarter Than You Think*. Plume.

Heal, Jane. 2003. *Mind, Reason and Imagination: Selected Essays in Philosophy of Mind and Language*. Cambridge University Press.

Heider, Fritz, and Marianne Simmel. 1944. "An Experimental Study of Apparent Behavior." *American Journal of Psychology* 57 (2): 243–59.

Helton, William S., and Nicole D. Helton. 2010. "Physical Size Matters in the Domestic Dog's (*Canis lupus familiaris*) Ability to Use Human Pointing Cues." *Behavioural Processes* 85 (1): 77–79.

Herzog, Hal. 2011. *Some We Love, Some We Hate, Some We Eat: Why It's So Hard to Think Straight about Animals*. HarperCollins.

Heyes, Cecilia M. 1998. "Theory of Mind in Nonhuman Primates." *Behavioral and Brain Sciences* 21 (1): 101–14, discussion 115–48.

———. 2012. "Grist and Mills: On the Cultural Origins of Cultural Learning." *Philosoph-*

ical Transactions of the Royal Society of London B: Biological Sciences 367 (1599): 2181–91.

Hibbard, Laura. 2011. "Dog of Fallen Navy SEAL, Officer Jon Tumilson, Refuses To Leave Casket." *Huffington Post*, August 24. https://www.huffpost.com/entry/dog-of-fallen-navy-seal-refuses-to-leave-casket_n_935707.

Horowitz, Alexandra. 2009. "Attention to Attention in Domestic Dog (*Canis familiaris*) Dyadic Play." *Animal Cognition* 12 (1): 107–18.

———. 2010. *Inside of a Dog: What Dogs See, Smell, and Know*. Scribner.

———. 2011. "Theory of Mind in Dogs?: Examining Method and Concept." *Learning and Behavior* 39 (4): 314–17.

———. 2016. *Being a Dog: Following the Dog into a World of Smell*. Scribner.

Horowitz, Alexandra C., and Marc Bekoff. 2007. "Naturalizing Anthropomorphism: Behavioral Prompts to Our Humanizing of Animals." *Anthrozoös* 20 (1): 23–35.

Howbert, J. Jeffry, Edward E. Patterson, S. Matt Stead, Ben Brinkmann, Vincent Vasoli, Daniel Crepeau, Charles H. Vite, et al. 2014. "Forecasting Seizures in Dogs with Naturally Occurring Epilepsy." *PLoS One* 9 (1): e81920.

Humphrey, N. 1977. Review of *The Question of Animal Awareness*, by D. R. Griffin. *Animal Behaviour* 25: 521–22.

Husserl, Edmund. 1960. *Cartesian Meditations*. Translated by Dorion Cairns. Martinus Nijhoff.

———. 1970. *The Crisis of European Sciences and Transcendental Phenomenology: An Introduction to Phenomenological Philosophy*. Translated by David Carr. Northwestern University Press.

Hutchins, Edwin. 2000. "Ecological Cognition and Cognitive Ecology." *Proceedings of the Human Factors and Ergonomics Society Annual Meeting* 44 (22): 566–69.

Hutto, Daniel D., and Erik Myin. 2012. *Radicalizing Enactivism: Basic Minds without Content*. MIT Press.

James, William. 1884. "What Is an Emotion?" *Mind: A Quarterly Review of Psychology and Philosophy* 9 (34): 188–205.

Jarvis, Pam. 2010. "'Born to Play': The Biocultural Roots of 'Rough and Tumble' Play, and Its Impact upon Young Children's Learning and Development." In Broadhead, Howard, and Wood 2010, 61–77.

Johnson, C. M. 2001. "Distributed Primate Cognition: A Review." *Animal Cognition* 3 (4): 167–83.

Johnson, Mark H., and John Morton. 1991. *Biology and Cognitive Development: The Case of Face Recognition*. Basil Blackwell.

Johnson, Mary C., and Karl L. Wuensch. 1994. "An Investigation of Habituation in the Jellyfish *Aurelia aurita*." *Behavioral and Neural Biology* 61 (1): 54–59.

Kaminski, Juliane, Josep Call, and Julia Fischer. 2004. "Word Learning in a Domestic Dog: Evidence for 'Fast Mapping.'" *Science* 304 (5677): 1682–83.

Kano, Fumihiro, and Josep Call. 2017. "Great Ape Social Attention." In *Evolution of the Brain, Cognition, and Emotion in Vertebrates*, edited by Shigeru Watanabe, Michel A. Hofman, and Toru Shimizu, 187–206. Springer Japan.

Karp, David A. 1996. *Speaking of Sadness: Depression, Disconnection, and the Meanings of Illness*. Oxford University Press.

Karpov, Yuriy V. 2005. *The Neo-Vygotskian Approach to Child Development*. Cambridge University Press.

Kato, Masaharu, and Ryoko Mugitani. 2015. "Pareidolia in Infants." *PLoS One* 10 (2): e0118539. doi:10.1371/journal.pone.0118539.

King, Barbara J. 2013. *How Animals Grieve*. University of Chicago Press.

Kirschner, S., and M. Tomasello. 2010. "Joint Music Making Promotes Prosocial Behavior in 4-Year-Old Children." *Evolution and Human Behavior* 31 (5): 354–64. https://www.ehbonline.org/article/S1090-5138(10)00046-2/fulltext.

Koehne, Svenja, Alexander Hatri, John T. Cacioppo, and Isabel Dziobek. 2016. "Perceived Interpersonal Synchrony Increases Empathy: Insights from Autism Spectrum Disorder." *Cognition* 146: 8–15.

Kraus, Cornelia, Cornelia van Waveren, and Franziska Huebner. 2014. "Distractible Dogs, Constant Cats? A Test of the Distraction Hypothesis in Two Domestic Species." *Animal Behaviour* 93 (July): 173–81.

Kraut, Robert. 1987. "Love De Re." *Midwest Studies in Philosophy* 10 (1): 413–30.

Krueger, Joel. 2014. "Varieties of Extended Emotions." *Phenomenology and the Cognitive Sciences* 13 (4): 533–55.

Kuba, Michael J., Ruth A. Byrne, Daniela V. Meisel, and Jennifer A. Mather. 2006. "When Do Octopuses Play? Effects of Repeated Testing, Object Type, Age, and Food Deprivation on Object Play in *Octopus vulgaris*." *Journal of Comparative Psychology* 120 (3): 184–90.

Legerstee, Maria, Diane Anderson, and Alliza Schaffer. 1998. "Five- and Eight-Month-Old Infants Recognize Their Faces and Voices as Familiar and Social Stimuli." *Child Development* 69 (1): 37–50.

Le Roux, Marieanna C., and Rene Kemp. 2009. "Effect of a Companion Dog on Depression and Anxiety Levels of Elderly Residents in a Long-Term Care Facility." *Psychogeriatrics: The Official Journal of the Japanese Psychogeriatric Society* 9 (1): 23–26.

Leslie, Alan M. 1987. "Pretense and Representation: The Origins of 'Theory of Mind.'" *Psychological Review* 94 (4): 412–26.

Levin, Jack, Arnold Arluke, and Leslie Irvine. 2017. "Are People More Disturbed by Dog or Human Suffering?: Influence of Victim's Species and Age." *Society and Animals* 25 (1): 1–16.

Lewandowski, Joseph. 2007. "Boxing: The Sweet Science of Constraints." *Journal of the Philosophy of Sport* 34 (1): 26–38.

Lidz, Jeffrey, Sandra Waxman, and Jennifer Freedman. 2003. "What Infants Know about Syntax but Couldn't Have Learned: Experimental Evidence for Syntactic Structure at 18 Months." *Cognition* 89 (3): B65–73.

Logan, Cheryl A. 1975. "Topographic Changes in Responding during Habituation to Water-Stream Simulation in Sea Anemones (*Anthopleura elegantissima*)." *Journal of Comparative and Physiological Psychology* 89 (2): 105–17.

Logan, Cheryl A., and Hall P. Beck. 1978. "Long-Term Retention of Habituation in the

Sea Anemone (*Anthopleura elegantissima*)." *Journal of Comparative and Physiological Psychology* 92 (5): 928–36.

Lucas, John. 2020. "Scout's Super Bowl Story Is a Viral Hit." *University of Wisconsin–Madison News*, February 3. https://news.wisc.edu/on-super-bowl-eve-scouts-story-is-a-viral-hit/.

Lurz, Robert W., ed. 2009. *The Philosophy of Animal Minds*. Cambridge University Press.

———. 2011. *Mindreading Animals: The Debate over What Animals Know about Other Minds*. MIT Press.

MacLean, Evan L., Esther Herrmann, Sunil Suchindran, and Brian Hare. 2017. "Individual Differences in Cooperative Communicative Skills Are More Similar between Dogs and Humans than Chimpanzees." *Animal Behaviour* 126 (April): 41–51.

Maguire, Eleanor A., Katherine Woollett, and Hugo J. Spiers. 2006. "London Taxi Drivers and Bus Drivers: A Structural MRI and Neuropsychological Analysis." *Hippocampus* 16 (12): 1091–1101.

Marchesini, Roberto. 2015. "Against Anthropocentrism: Non-Human Otherness and the Post-Human Project." *NanoEthics* 9 (1): 75–84.

Marr, David. (1982) 2010. *Vision: A Computational Investigation into the Human Representation and Processing of Visual Information*. MIT Press.

Martin, Alia, and Laurie R. Santos. 2016. "What Cognitive Representations Support Primate Theory of Mind?" *Trends in Cognitive Sciences* 20 (5): 375–82.

Marzluff, John M., and Tony Angell. 2005. *In the Company of Crows and Ravens*. Yale University Press.

———. 2012. *Gifts of the Crow: How Perception, Emotion, and Thought Allow Smart Birds to Behave Like Humans*. Free Press / Simon and Schuster.

Marzluff, John M., Robert Miyaoka, Satoshi Minoshima, and Donna J. Cross. 2012. "Brain Imaging Reveals Neuronal Circuitry Underlying the Crow's Perception of Human Faces." *Proceedings of the National Academy of Sciences of the United States of America* 109 (39): 15912–17.

Marzluff, John M., Jeff Walls, Heather N. Cornell, John C. Withey, and David P. Craig. 2010. "Lasting Recognition of Threatening People by Wild American Crows." *Animal Behaviour* 79 (3): 699–707.

Maynard-Smith, John, and David Harper. 2004. *Animal Signals*. Oxford University Press.

McComb, Karen, Cynthia J. Moss, Soila Sayialel, and Lucy Baker. 2000. "Unusually Extensive Networks of Vocal Recognition in African Elephants." *Animal Behaviour* 59 (6): 1103–9.

McConnell, Patricia B. 2006. *For the Love of a Dog: Understanding Emotion in You and Your Best Friend*. Ballantine.

McVean, Ada. 2019. "Leafcutter Ants Are Farmers Who Grow Fungi." McGill Office for Science and Society. https://www.mcgill.ca/oss/article/did-you-know/did-you-know-leafcutter-ants-are-farmers-who-grow-fungi.

Mead, George Herbert. 1934. *Mind, Self and Society from the Standpoint of a Social Behaviorist*. Edited by Charles W. Morris. University of Chicago Press.

Meltzoff, Andrew N. 2007. "'Like Me': A Foundation for Social Cognition." *Developmental Science* 10 (1): 126–34.
Meltzoff, Andrew N., and M. Keith Moore. 1977. "Imitation of Facial and Manual Gestures by Human Neonates." *Science* 198 (4312): 75–78.
Menary, Richard. 2006. "Attacking the Bounds of Cognition." *Philosophical Psychology* 19 (3): 329–44.
———. 2010. *The Extended Mind*. MIT Press.
Merritt, Michele. 2013. "Instituting Impairment: Extended Cognition and the Construction of Female Sexual Dysfunction." *Cognitive Systems Research* 25–26 (December): 47–53.
———. 2015a. "Dismantling Standard Cognitive Science: It's Time the Dog Has Its Day." *Biology and Philosophy* 30 (6): 811–29.
———. 2015b. "Thinking-Is-Moving: Dance, Agency, and a Radically Enactive Mind." *Phenomenology and the Cognitive Sciences* 14 (1): 95–110.
Miklósi, Ádám. 2014. *Dog Behaviour, Evolution, and Cognition*. Oxford University Press.
Miklósi, Ádám, and Enikő Kubinyi. 2016. "Current Trends in Canine Problem-Solving and Cognition." *Current Directions in Psychological Science* 25 (5): 300–306.
Miklósi, Ádám, Rob Polgárdi, József Topál, and Vilmos Csányi. 1998. "Use of Experimenter-Given Cues in Dogs." *Animal Cognition* 1 (2): 113–21.
Miklósi, Ádám, and Krisztina Soproni. 2006. "A Comparative Analysis of Animals' Understanding of the Human Pointing Gesture." *Animal Cognition* 9 (2): 81–93.
Miklósi, Ádám, József Topál, and Vilmos Csányi. 2004. "Comparative Social Cognition: What Can Dogs Teach Us?" *Animal Behaviour* 67 (6): 995–1004.
Mitchell, Robert W. 2001. "Americans' Talk to Dogs: Similarities and Differences with Talk to Infants." *Research on Language and Social Interaction* 34 (2): 183–210.
———. 2004. "Controlling the Dog, Pretending to Have a Conversation, or Just Being Friendly?: Influences of Sex and Familiarity on Americans' Talk to Dogs during Play." *Interaction Studies* 5 (1): 99–129.
———. 2015. "Creativity in the Interaction: The Case of Dog-Human Play." In *Animal Creativity and Innovation*, edited by Allison Kaufman and James C. Kaufman, 31–44. Elsevier.
Mitchell, Robert W., Emily Reed, and Lyndsey Alexander. 2018. "Functions of Pointing by Humans, and Dogs' Responses, During Dog-Human Play between Familiar and Unfamiliar Players." *Animal Behavior and Cognition* 5 (2): 181–200.
Mitchell, Robert W., and Nicholas S. Thompson, eds. 1986a. *Deception: Perspectives on Human and Nonhuman Deceit*. SUNY Press.
———. 1986b. "Deception in Play between Dogs and People." In Mitchell and Thompson 1986a, 193–204.
———. 1990. "The Effects of Familiarity on Dog-Human Play." *Anthrozoös* 4 (1): 24–43.
———. 1991. "Projects, Routines, and Enticements in Dog-Human Play." *Perspectives in Ethology* 9: 189–216.
Montero, Barbara Gail. 2016. *Thought in Action: Expertise and the Conscious Mind*. Oxford University Press.
Moore, Derek G., R. Peter Hobson, and Anthony Lee. 1997. "Components of Person Per-

ception: An Investigation with Autistic, Non-Autistic Retarded and Typically Developing Children and Adolescents." *British Journal of Developmental Psychology* 15 (4): 401–23.

Morgan, C. Lloyd. (1894) 1977. *An Introduction to Comparative Psychology; "The Limits of Animal Intelligence."* Repr., University Publications of America.

Moyles, Janet R., ed. 2014. *The Excellence of Play*. 4th ed. McGraw-Hill Education (UK).

Mubanga, Mwenya, Liisa Byberg, Christoph Nowak, Agneta Egenvall, Patrik K. Magnusson, Erik Ingelsson, and Tove Fall. 2017. "Dog Ownership and the Risk of Cardiovascular Disease and Death—a Nationwide Cohort Study." *Scientific Reports* 7 (1): 15821.

Nagel, Thomas. 1974. "What Is It Like to Be a Bat?" *Philosophical Review* 83 (4): 435–50.

Newell, Allen, and Herbert A. Simon. 1976. "Computer Science As Empirical Inquiry: Symbols and Search." *Communications of the ACM* 19 (3): 113–26.

Newen, Albert, Leon De Bruin, and Shaun Gallagher, eds. 2018. *The Oxford Handbook of 4E Cognition*. Oxford University Press.

Newman, Lesléa. 2004. *Hachiko Waits*. Henry Holt.

Nichols, Shaun, and Stephen P. Stich. 2003. *Mindreading: An Integrated Account of Pretence, Self-Awareness, and Understanding Other Minds*. Clarendon Press.

Noë, Alva. 2004. *Action in Perception*. MIT Press.

———. 2009. *Out of Our Heads: Why You Are Not Your Brain, and Other Lessons from the Biology of Consciousness*. Hill and Wang.

Nussbaum, Martha C. 2001. *Upheavals of Thought: The Intelligence of Emotions*. Cambridge University Press.

Odendaal, J. S. J., and R. A. Meintjes. 2003. "Neurophysiological Correlates of Affiliative Behaviour between Humans and Dogs." *Veterinary Journal* 165 (3): 296–301.

Oliveira, Ana Flora Sarti, André Oliveira Rossi, Luana Finocchiaro Romualdo Silva, Michele Correa Lau, and Rodrigo Egydio Barreto. 2010. "Play Behaviour in Nonhuman Animals and the Animal Welfare Issue." *Journal of Ethology* 28 (1): 1–5.

O'Regan, J. Kevin, and Alva Noë. 2001. "A Sensorimotor Account of Vision and Visual Consciousness." *Behavioral and Brain Sciences* 24 (5): 939–73, discussion 973–1031.

Overgaard, Søren. 2006. "The Problem of Other Minds: Wittgenstein's Phenomenological Perspective." *Phenomenology and the Cognitive Sciences* 5 (1): 53–73.

Palagi, Elisabetta, Velia Nicotra, and Giada Cordoni. 2015. "Rapid Mimicry and Emotional Contagion in Domestic Dogs." *Royal Society Open Science* 2 (12): 150505.

Payne, Elyssa, Pauleen C. Bennett, and Paul D. McGreevy. 2015. "Current Perspectives on Attachment and Bonding in the Dog-Human Dyad." *Psychology Research and Behavior Management* 8 (February): 71–79.

Pellegrini, Anthony D. 2009. *The Role of Play in Human Development*. Oxford University Press.

Pellegrini, Anthony D., and Kathy Gustafson. 2005. "Boys' and Girls' Uses of Objects for Exploration, Play, and Tools in Early Childhood." In *The Nature of Play: Great Apes and Humans*, edited by Anthony D. Pellegrini and Peter K. Smith, 113–35. Guilford Press.

Pellegrini, A. D., and Peter K. Smith. 1998. "Physical Activity Play: The Nature and Function of a Neglected Aspect of Playing." *Child Development* 69 (3): 577–98.

Penn, Derek C., and Daniel J. Povinelli. 2007a. "Causal Cognition in Human and Nonhuman Animals: A Comparative, Critical Review." *Annual Review of Psychology* 58: 97–118.

———. 2007b. "On the Lack of Evidence That Non-Human Animals Possess Anything Remotely Resembling a 'Theory of Mind.'" *Philosophical Transactions of the Royal Society of London B: Biological Sciences* 362 (1480): 731–44.

Pfungst, Oskar. 1911. *Clever Hans (the Horse of Mr. Von Osten): A Contribution to Experimental Animal and Human Psychology*. Henry Holt.

Pilley, John W., and Alliston K. Reid. 2011. "Border Collie Comprehends Object Names as Verbal Referents." *Behavioural Processes* 86 (2): 184–95.

Plato. 1997. *Complete Works*. Edited by John M. Cooper and D. S. Hutchinson. Hackett.

Plotnik, Joshua M., Frans B. M. de Waal, and Diana Reiss. 2006. "Self-Recognition in an Asian Elephant." *Proceedings of the National Academy of Sciences of the United States of America* 103 (45): 17053–57.

Pollick, Amy S., and Frans B. M. de Waal. 2007. "Ape Gestures and Language Evolution." *Proceedings of the National Academy of Sciences of the United States of America* 104 (19): 8184–89.

Porges, Stephen W. 2011. *Polyvagal Theory: Neurophysiological Foundations of Emotions, Attachment, Communication, and Self-Regulation*. Norton.

Povinelli, Daniel J., and Timothy J. Eddy. 1996. "What Young Chimpanzees Know about Seeing." *Monographs of the Society for Research in Child Development* 61 (3): i–vi, 1–152, discussion 153–91.

Povinelli, Daniel J., and Jennifer Vonk. 2003. "Chimpanzee Minds: Suspiciously Human?" *Trends in Cognitive Sciences* 7 (4): 157–60.

Power, Thomas G. 1999. *Play and Exploration in Children and Animals*. Psychology Press.

Premack, David, and Guy Woodruff. 1978. "Does the Chimpanzee Have a Theory of Mind?" *Behavioral and Brain Sciences* 1 (4): 515–26.

Prinz, Jesse J. 2004. *Gut Reactions: A Perceptual Theory of Emotion*. Oxford University Press.

Proops, Leanne, Kate Grounds, Amy Victoria Smith, and Karen McComb. 2018. "Animals Remember Previous Facial Expressions That Specific Humans Have Exhibited." *Current Biology* 28 (9): 1428–32.e4.

Putnam, Hilary. 1975. "'The Meaning of 'Meaning.'" In *Language, Mind, and Knowledge*, edited by Keith Gunderson, 131–93. University of Minnesota Press.

Racca, Anaïs, Kun Guo, Kerstin Meints, and Daniel S. Mills. 2012. "Reading Faces: Differential Lateral Gaze Bias in Processing Canine and Human Facial Expressions in Dogs and 4-Year-Old Children." *PLoS One* 7 (4): e36076.

Rajmohan, V., and E. Mohandas. 2007. "Mirror Neuron System." *Indian Journal of Psychiatry* 49 (1): 66–69.

Ratcliffe, Matthew. 2010. "Depression, Guilt and Emotional Depth." *Inquiry* 53 (6): 602–26.

Reddy, Vasudevi, and Colwyn Trevarthen. 2004. "What We Learn about Babies from Engaging Their Emotions." *Zero to Three* 24 (3): 9–15.

Resende, Briseida Dôgo de, and Eduardo B. Ottoni. 2002. "Play and Tool Use Learning in Tufted Capuchin Monkeys (*Cebus apella*)." *Estudos de Psicologia* (Natal) 7 (1): 173–80.

Rivas, Jesús, and Gordon M. Burghardt. 2002. "Crotalomorphism: A Metaphor for Understanding Anthropomorphism by Omission." In Bekoff, Allen, and Burghardt 2002, 9–17.

Rizzolatti, Giacomo, and Laila Craighero. 2004. "The Mirror-Neuron System." *Annual Review of Neuroscience* 27 (1): 169–92.

Rizzolatti, Giacomo, Luciano Fadiga, Vittorio Gallese, and Leonardo Fogassi. 1996. "Premotor Cortex and the Recognition of Motor Actions." *Cognitive Brain Research* 3 (2): 131–41.

Rizzolatti, Giacomo, and Giuseppi Luppino. 2001. "The Cortical Motor System." *Neuron* 31 (6): 889–901.

Rodionova, Elena, Alexander Minor, Klim Sulimov, and Galina Kogun. 2009. "Dogs Are Able to Recognize Insect Individuals by Odour." In *Chemical Senses* 34 (3):E64–E64.

Rosenthal, David M. 1993. "Higher-Order Thoughts and the Appendage Theory of Consciousness." *Philosophical Psychology* 6 (2): 155–66.

———. 2004. "Varieties of Higher-Order Theory." *Advances in Consciousness Research* 56: 17–44.

Rushforth, Norman B., Ivan T. Krohn, and Lee K. Brown. 1964. "Behavior in *Hydra pirardi*: Inhibition of the Contraction Responses of *Hydra pirardi*." *Science* 145 (3632): 602–4.

Ryall, Emily, Wendy Russell, and Malcolm MacLean, eds. 2013. *The Philosophy of Play*. Routledge.

Ryle, Gilbert. 1968. *The Thinking of Thoughts*. University of Saskatchewan.

Saffran, Jenny R., Ann Senghas, and John C. Trueswell. 2001. "The Acquisition of Language by Children." *Proceedings of the National Academy of Sciences of the United States of America* 98 (23): 12874–75.

Samhita, Laasya, and Hans J. Gross. 2013. "The 'Clever Hans Phenomenon' Revisited." *Communicative and Integrative Biology* 6 (6): e2722.

Santuzzi, Alecia M., Pamela R. Waltz, Lisa M. Finkelstein, and Deborah E. Rupp. 2014. "Invisible Disabilities: Unique Challenges for Employees and Organizations." *Industrial and Organizational Psychology* 7 (2): 204–19.

Scheve, Christian von, and Mikko Salmela, eds. 2014. *Collective Emotions: Perspectives from Psychology, Philosophy, and Sociology*. Oxford University Press.

Scholl, Brian J., and Alan M. Leslie. 1999. "Modularity, Development and 'Theory of Mind.'" *Mind and Language* 14 (1): 131–53.

Schore, Allan N. 2000. "Attachment and the Regulation of the Right Brain." *Attachment and Human Development* 2 (1): 23–47.

Searle, John R. 1983. *Intentionality: An Essay in the Philosophy of Mind*. Cambridge University Press.

Sebeok, Thomas A. 1976. *Contributions to the Doctrine of Signs*. Indiana University.

Seemann, Axel, ed. 2011. *Joint Attention: New Developments in Psychology, Philosophy of Mind, and Social Neuroscience*. MIT Press.

Sewall, K. 2015. "The Girl Who Gets Gifts from Birds." *BBC News Magazine*, February 25. https://www.bbc.com/news/magazine-31604026.
Shik, Jonathan Zvi, Winnie Rytter, Xavier Arnan, and Anders Michelsen. 2018. "Disentangling Nutritional Pathways Linking Leafcutter Ants and Their Co-evolved Fungal Symbionts Using Stable Isotopes." *Ecology* 99 (9): 1999–2009. doi:10.1002/ecy.2431.
Shipman, Pat. 2015. *The Invaders: How Humans and Their Dogs Drove Neanderthals to Extinction*. Harvard University Press.
Sieczkowski, Cavan. 2013. "Baby Elephant Cries for 5 Hours After Mom Attacks, Rejects Him." *Huffington Post*, September 13. https://www.huffpost.com/entry/baby-elephant-cries_n_3920685.
Simpson, M. J. A. 1976. "The Study of Animal Play." In *Growing Points in Ethology*, edited by P. P. G. Bateson and R. A. Hinde, 385–400. Cambridge University Press.
Skinner, B. F. 1953. *Science and Human Behavior*. Macmillan.
———. 1974. *About Behaviorism*. Knopf.
Slaby, Jan. 2014. "Emotions and the Extended Mind." In Scheve and Salmela 2014, 32–46.
Smith, Amy Victoria, Leanne Proops, Kate Grounds, Jennifer Wathan, and Karen McComb. 2016. "Functionally Relevant Responses to Human Facial Expressions of Emotion in the Domestic Horse (*Equus caballus*)." *Biology Letters* 12 (2): 20150907.
Smith, Don. 1993. "Drowning Puppy." YouTube. https://www.youtube.com/watch?v=_PR5moWpa5Q.
Sokolov, V. E., V. I. Krutova, K. T. Sulimov, and E. P. Zinkevich. 1997. "Olfactory Identification of Individual Species Specificity in Dogs" (in Russian). *Doklady Akademii Nauk* (Proceedings of the Academy of Sciences) 356: 716–18.
Solomon, Robert C. 1973. "Emotions and Choice." *Review of Metaphysics* 27 (1): 20–41.
Somerville, James. 1989. "Making Out the Signatures: Reid's Account of the Knowledge of Other Minds." In *The Philosophy of Thomas Reid*, edited by Melvin Dalgamo and Eric Matthews, 249–73. Kluwer.
Soproni, Krisztina, Ádám Miklósi, József Topál, and Vilmos Csányi. 2001. "Comprehension of Human Communicative Signs in Pet Dogs (*Canis familiaris*)." *Journal of Comparative Psychology* 115 (2): 122–26.
Sorce, James F., Robert N. Emde, Joseph J. Campos, and Mary D. Klinnert. 1985. "Maternal Emotional Signaling: Its Effect on the Visual Cliff Behavior of 1-Year-Olds." *Developmental Psychology* 21 (1): 195–200.
Spaulding, Shannon. 2015. "On Direct Social Perception." *Consciousness and Cognition* 36 (November): 472–82.
Sternberg, Robert J., and Todd I. Lubart. 1999. "The Concept of Creativity: Prospects and Paradigms." In *Handbook of Creativity*, edited by Robert J. Sternberg, 3–15. Cambridge University Press.
Strong, Val, Stephen W. Brown, and Robin Walker. 1999. "Seizure-Alert Dogs—fact or Fiction?" *Seizure: The Journal of the British Epilepsy Association* 8 (1): 62–65.
Surtees, Andrew D. R., Stephen A. Butterfill, and Ian A. Apperly. 2011. "Direct and Indi-

rect Measures of Level-2 Perspective-Taking in Children and Adults." *British Journal of Developmental Psychology* 30 (1): 75–86.

Sylva, Kathy, Jerome S. Bruner, and Paul Genova. 1976. "The Role of Play in the Problem-Solving of Children 3–5 Years Old." In *Play: Its Role in Development and Evolution*, edited by Jerome S. Bruner, Alison Jolly, and Kathy Sylva, 244–57. Basic Books.

Thagard, Paul. 2018. "Cognitive Science." In *The Stanford Encyclopedia of Philosophy*, evolving online resource edited by Edward N. Zalta. https://plato.stanford.edu/entries/cognitive-science/.

Thompson, Evan. 2007. *Mind in Life: Biology, Phenomenology, and the Sciences of Mind*. Belknap Press / Harvard University Press.

Tibbles, Kevin. 2020. "Why This Man Took Out a $6 Million Super Bowl Ad for His Dog." *NBC Nightly News*, January 30. https://www.nbcnews.com/nightly-news/video/why-this-man-took-out-a-6-million-super-bowl-ad-for-his-dog-77830725929.

Timberlake, William, and Andrew R. Delamater. 1991. "Humility, Science, and Ethological Behaviorism." *Behavior Analyst / MABA* 14 (1): 37–41.

Tinbergen, N. 1951. *The Study of Instinct*. Clarendon Press.

———. 1963. "On Aims and Methods of Ethology." *Zeitschrift für Tierpsychologie* 20 (4): 410–33.

Tollefsen, Deborah. 2005. "Let's Pretend! Children and Joint Action." *Philosophy of the Social Sciences* 35 (1): 75–97.

Tomonaga, Masaki, Yuka Uwano, Sato Ogura, Hyangsun Chin, Masahiro Dozaki, and Toyoshi Saito. 2015. "Which Person Is My Trainer? Spontaneous Visual Discrimination of Human Individuals by Bottlenose Dolphins (*Tursiops truncatus*)." *SpringerPlus* 4 (July): 352.

Trevarthen, Colwyn. 1979. "Communication and Cooperation in Early Infancy: A Description of Primary Intersubjectivity." In *Before Speech: The Beginning of Interpersonal Communication*, edited by Margaret Bullowa, 530–71. Cambridge University Press.

———. 1999. "Musicality and the Intrinsic Motive Pulse: Evidence from Human Psychobiology and Infant Communication." *Musicae Scientiae: The Journal of the European Society for the Cognitive Sciences of Music* 3 (1_suppl): 155–215.

Triantafyllou, Michael S., and George S. Triantafyllou. 1995. "An Efficient Swimming Machine." *Scientific American* 272 (3): 64–70.

Tronick, Edward, Heidelise Als, Lauren Adamson, Susan Wise, and T. Berry Brazelton. 1978. "The Infant's Response to Entrapment between Contradictory Messages in Face-to-Face Interaction." *Journal of the American Academy of Child Psychiatry* 17 (1): 1–13.

Trut, Lyudmila N. 1999. "Early Canid Domestication: The Farm-Fox Experiment: Foxes Bred for Tamability in a 40-Year Experiment Exhibit Remarkable Transformations That Suggest an Interplay between Behavioral Genetics and Development." *American Scientist* 87 (2): 160–69.

Trut, Lyudmila, Irina Oskina, and Anastasiya Kharlamova. 2009. "Animal Evolution during Domestication: The Domesticated Fox as a Model." *BioEssays: News and Reviews in Molecular, Cellular and Developmental Biology* 31 (3): 349–60.

Trut, L. N., I. Z. Plyusnina, and I. N. Oskina. 2004. "An Experiment on Fox Domestication and Debatable Issues of Evolution of the Dog" (in Russian). *Genetika* 40 (6): 644–55.

Turner, Pamela S. 2004. *Hachiko: The True Story of a Loyal Dog*. Houghton Mifflin.

Udell, Monique A. R. 2015. "When Dogs Look Back: Inhibition of Independent Problem-Solving Behaviour in Domestic Dogs (*Canis lupus familiaris*) Compared with Wolves (*Canis lupus*)." *Biology Letters* 11 (9): 20150489.

Udell, Monique A. R., Nicole R. Dorey, and Clive D. L. Wynne. 2008. "Wolves Outperform Dogs in Following Human Social Cues." *Animal Behaviour* 76 (6): 1767–73.

———. 2010a. "The Performance of Stray Dogs (*Canis familiaris*) Living in a Shelter on Human-Guided Object-Choice Tasks." *Animal Behaviour* 79 (3): 717–25.

———. 2010b. "What Did Domestication Do to Dogs? A New Account of Dogs' Sensitivity to Human Actions." *Biological Reviews of the Cambridge Philosophical Society* 85 (2): 327–45.

Udell, Monique A. R., Margaret Ewald, Nicole R. Dorey, and Clive D. L. Wynne. 2014. "Exploring Breed Differences in Dogs (*Canis familiaris*): Does Exaggeration or Inhibition of Predatory Response Predict Performance on Human-Guided Tasks?" *Animal Behaviour* 89 (March): 99–105.

Varela, Francisco J., Evan Thompson, and Eleanor Rosch. 1991. *The Embodied Mind: Cognitive Science and Human Experience*. MIT Press.

Varga, Somogy, and Joel Krueger. 2013. "Background Emotions, Proximity and Distributed Emotion Regulation." *Review of Philosophy and Psychology* 4 (2): 271–92.

Virani, Salim S., A. Nasser Khan, Cesar E. Mendoza, Alexandre C. Ferreira, and Eduardo de Marchena. 2007. "Takotsubo Cardiomyopathy, or Broken-Heart Syndrome." *Texas Heart Institute Journal* 34 (1): 76–79.

Vonk, Jennifer, and Daniel J. Povinelli. 2011. "Social and Physical Reasoning in Human-Reared Chimpanzees: Preliminary Studies." In *Perception, Causation, and Objectivity*, edited by Johannes Roessler, Hemdat Lerman, and Naomi Eilan, 342–67. Oxford University Press.

von Uexküll, Jakob. (1934) 1957. "A Stroll through the Worlds of Animals and Men." In *Instinctive Behavior: The Development of a Modern Concept*, edited and translated by Claire H. Schiller, 5–80. International University Press.

Vygotsky, Lev Semenovich. 1986. *Thought and Language*. Edited and translated by Eugenia Hanfmann and Gertrude Vakar, revised by Alex Kazulin. MIT Press.

Waal, Frans de. 1996. *Good Natured: The Origins of Right and Wrong in Humans and Other Animals*. Harvard University Press.

———. 2006. *Primates and Philosophers: How Morality Evolved*. Princeton University Press.

Walker, A. S. 1982. "Intermodal Perception of Expressive Behaviors by Human Infants." *Journal of Experimental Child Psychology* 33 (3): 514–35.

Warren, Cat. 2013. *What the Dog Knows: The Science and Wonder of Working Dogs*. Simon and Schuster.

"WeatherTech Super Bowl ad featuring Scout." 2020. YouTube, February 2. https://www.youtube.com/watch?v=Fi2WwRJDii.

Westra, Evan. 2017. "Spontaneous Mindreading: A Problem for the Two-Systems Account." *Synthese* 194 (11): 4559–81.

Whitebread, David. 2017. "Prioritizing Play." In *EarthEd: Rethinking Education on a Changing Planet*, edited by Erik Assadourian and Lisa Mastny, 107–16. Island Press.

Willis, Carolyn M., Susannah M. Church, Claire M. Guest, W. Andrew Cook, Noel McCarthy, Anthea J. Bransbury, Martin R. T. Church, and John C. T. Church. 2004. "Olfactory Detection of Human Bladder Cancer by Dogs: Proof of Principle Study." *BMJ* 329 (7468): 712.

Winsler, Adam, and Jack Naglieri. 2003. "Overt and Covert Verbal Problem-Solving Strategies: Developmental Trends in Use, Awareness, and Relations with Task Performance in Children Aged 5 to 17." *Child Development* 74 (3): 659–78.

Wittgenstein, Ludwig. (1953) 2009. *Philosophical Investigations*. 4th ed. German text, with translation by G. E. M. Anscombe, revised by P. M. S. Hacker and Joachim Schulte. Wiley-Blackwell.

Wittstein, Ilan S. 2007. "The Broken Heart Syndrome." *Cleveland Clinic Journal of Medicine* 74 Suppl 1 (February): S17–22.

Wobber, Victoria, Brian Hare, Janice Koler-Matznick, Richard W. Wrangham, and Michael Tomasello. 2009. "Breed Differences in Domestic Dogs' (*Canis familiaris*) Comprehension of Human Communicative Signals." *Interaction Studies* 10 (2): 206–24.

Wollerton, Megan. 2019. "Revenge of the Dogs: Sony Aibo Does Not Impress Your Furry Friends." CNET, June 27. https://www.cnet.com/news/dont-bank-on-sonys-aibo-robot-dog-being-your-dogs-best-friend/.

Wynne, Clive D. L. 2007. "What Are Animals? Why Anthropomorphism Is Still Not a Scientific Approach to Behavior." *Comparative Cognition and Behavior Reviews* 2 (1): 125–35. https://courses.washington.edu/anmind/Wynne-anthropomorphism-CCBR2007.pdf.

Wynne, Clive D. L., Monique A. R. Udell, and Kathryn A. Lord. 2008. "Ontogeny's Impacts on Human-Dog Communication." *Animal Behaviour* 76 (4): e1–4.

Young, Michael J., and Ethan Scheinberg. 2017. "The Rise of Crowdfunding for Medical Care: Promises and Perils." *JAMA* 317 (16): 1623–24. doi:10.1001/jama.2017.3078.

Zaine, Isabela, Camila Domeniconi, and Clive D. L. Wynne. 2015. "The Ontogeny of Human Point Following in Dogs: When Younger Dogs Outperform Older." *Behavioural Processes* 119 (October): 76–85.

INDEX

Adamson, Lauren, 45
affect: attunement and, 62–63, 74, 104, 139–40, 143, 148–49; background emotion and, 46–48, 61; cognition and, 3, 5, 15, 35, 37, 44–47, 66, 74–75, 101–2; dog perception of human, 59; of dogs, 52, 61–62, 65; ecological, 20, 48; grief and, 49, 64–65, 164n15; internalist-externalist debate, 63–64, 65; phenomenal coupling, 49
Agamben, Giorgio, 14
agility sport, 58–59, 98, 118–19
AIBO (Sony robotic dog), 156–57, 166n3
Allen, Colin, 2, 13, 24, 60–61
Allen, Micah, 9, 126, 139–43
Andrews, Kristin, 79–81, 112, 150–51, 153; *Do Apes Read Minds?*, 89–90
Animal Thinking (Griffin), 11
anthropectomy, 24, 112
anthropocentrism, 24, 30, 72
anthropomorphism, 2, 13, 18, 25, 30, 98–99, 133–34 (*see also* critical anthropomorphism); by omission, 59, 105; sentimental, 153
ants, 131–32, 134, 137, 151
apes. *See* primates, nonhuman
apparatuses of encounters, 121–22, 145
Apperly, Ian, 93
appraisal theory, 44
Aristotle, 100–101

autopoiesis, 8, 99, 134, 138–40, 144; enactivism, incompleteness of theory regarding, 123–24, 125–26, 155
awareness relations, 77, 86–87

Baron-Cohen, Simon, 69, 74, 91
Barrett, Lisa, 94–95
bats, 27
Bayesian inference, 135, 140
behaviorism, 60
Bekoff, Marc, 2, 13, 24, 60–61, 104, 108, 114, 156
belief (intentional state), 72, 115–16
Belyayev, Dmitry, 53
Berk, Laura, 101–2, 103
Bermúdez, José, 78–79, 82, 88–89
Berns, Gregory, 59
birds, 104, 107; corvids, 84; parrots, 65, 122
Bordenstein, Seth, 8
boxing (human sport), 129
Braitman, Laurel, 59, 61–62
breeders (of dogs), 120–21
Brooks, Rodney, 95
bucket head tests, 80–81, 84
Burghardt, Gordon, 25–26, 30, 105–7, 126, 146; *The Genesis of Animal Play*, 105
Butterfill, Stephen, 93
Buytendijk, F. J. J., 111
Byers, John, 114

Cartesian philosophy, 36, 38, 43–44, 51–52; Réné Descartes, 27–28, 38
Chalmers, David, 47
Chaser (border collie), 2–3, 39, 52, 163n6
Chemero, Anthony, 72
Cheng, Ken, 134, 146–47, 151, 156
Chevalier-Skolnikoff, Suzanne, 114
Chomsky, Noam, 79
Chris and Hercules (dog-human play pair), 128–29, 141
Churchland, Paul, 116
Clark, Andy, 8–9, 47, 95, 126, 130, 133, 135–39, 140, 142–43, 144, 148–51; *Supersizing the Mind*, 138; *Surfing Uncertainty*, 135–38
classical ethology, 11
Clever Hans Phenomenon, 16–18, 22
Cnidaria (siphonophores), 8, 145–48
coactive cognition, 3, 6, 10, 35–36, 46, 56; affect and, 67; Cartesian critique and, 55; human-dog, 59, 61, 63–64, 96, 126; pro-social engagement and, 53–54; social sense-making and, 60; thinking-in-playing, 111, 122. *See also* play
Coelho, Luis Pedro, 132
cognitive ethology, 4, 11–13, 15, 60, 107; problem of domestication and, 19–22, 85; subjective phenomenology, 46, 51
cognitive scaffolding, 5
cognitivism, theory of mentality, 36, 43–44
collaborative cognition. *See* coactive cognition
computational account of the mind, 36, 43–44
constraint theory, 128–29
conversation of gestures, 99, 117–20
creativity. *See* play, creativity and
critical anthropomorphism, 4–5, 6, 25–27, 30, 33–34, 41, 105
critical theory, 111
Cronin, Helena, 13
Currie, Gregory, 57
cynomorphism, 31, 33, 157–58

dance, 57, 130, 166n2
Darwin (dog of Michele Merritt), 54, 106, 119
Darwin, Charles, 12
De Jaegher, Hanne, 57, 130
de Waal, Frans, 26, 30
deception, 110, 112–15, 117, 127–28
Delamater, Andrew, 26
Dempster, M. Beth, 123

Dennett, Daniel, 29–30, 33–34, 110, 115–16, 120–21, 136
Descartes, Réné, 27–28, 38; *Discourse*, 38–39. *See also* Cartesian philosophy
design stance, 34
Despret, Vinciane, 99, 117, 120–21, 123–24, 131, 144–45
Dewey, John, 20, 140–41
Di Paolo, Ezequiel, 57, 130, 139
Discourse (Descartes), 38–39
Do Apes Read Minds? (Andrews), 89–90
Dognition, 15–16, 38, 162n6
dogs, street, 16, 19, 31, 37, 41, 63, 161n4
dolphins, 10, 19–20, 39, 157, 163n4
domestication, 19–22, 85
Dorey, Nicole, 83
dyadic pairings. *See* coactive cognition

ecological cognition, 20, 48
Elster, Jon, 129
Embodied Mind, The (Varela, Thompson, and Rosch), 47
emotion. *See* affect
emotional support animals (ESAs), 158–60, 161n1
enactivist approach to cognition, 5–6, 49–50, 61, 98–99, 139, 155–56; Interaction Theory and, 67; reading, 56
escalating reciprocity, 128, 130, 138

face, ability to read, 18, 40–41, 45, 73–74, 162n9
fallacy, coupling-constitution, 139
fallacy, nominalist, 26, 82
False Belief Test, 69–70, 77
Family Dog Lab, 16, 38
feminism, 111–12
Fodor, Jerry, 28
folk psychology, 1, 13, 60–61, 89–91, 93–94, 115–16
Fonagy, Peter, 48
fox, silver, 52–53
Frick, Janet, 45

Gácsi, Márta, 83
Gadamer, Hans-Georg, 140
Gallagher, Shaun, 9, 29, 67, 72–73, 75–76, 98, 126, 132–33, 139–43, 148–51; *How the Body Shapes the Mind*, 73
Gallese, Vittorio, 69

Geertz, Clifford, 114
Genesis of Animal Play, The (Burghardt), 105
Gibson, James, 48
Godfrey-Smith, Peter, 30–31
Goldman, Alvin, 69
Gopnik, Alison, 69, 74
Gordon, Robert, 71
Griffin, Donald, 11–13, 15; *Animal Thinking*, 11; *The Question of Animal Awareness*, 11
Grimes, John, 30

Haddock, Stephen, 147
Haraway, Donna, 7–8, 57–58, 99, 111–12, 117–20, 129, 155, 157, 160; *Staying with the Trouble*, 122–24; *When Species Meet*, 118–19
Hare, Brian, 15–16, 38, 41, 54, 59, 78, 80–81, 83, 95, 97
Harper, David, 118
Heal, Jane, 69
Heider, Fritz, 75
Helton, Nicole, 83
Helton, William, 83
Hercules and Chris (dog-human play pair), 128–29, 141
hermeneutics, 139–41, 142–43, 151
heterophenomenology, 29–30, 33
Heyes, Cecilia, 80–81, 82, 84–85, 133
holobiont theory, 8, 99, 125, 131–32, 155
Horowitz, Alexandra, 26–27, 31, 33, 59, 86–87, 90–92, 93, 156
How the Body Shapes the Mind (Gallagher), 73
Humphrey, N., 13
Huss, Brian, 112
Husserl, Edmund, 29
Hutto, Daniel, 48, 95

infants, human, 74–75, 142
intelligence, 22–23
intentional stance, 34, 44, 110–11
Interaction Theory, 67, 69, 72, 75–76, 86, 97
interactive cognition. *See* coactive cognition
intersubjectivity, 45–46, 131, 151–52 (*see also* Interaction Theory); primary intersubjectivity, 6–7, 67, 72–73, 75–76

Johnson, C. M., 95
joint action, 150
Jung, Carl, 102

Karp, David, 48
Karpov, Yuriy, 103
Kirschner, S., 103
Kish, Daniel, 27
Kraus, Cornelia, 83

Laws (Plato), 100
left-gaze bias, 40–41
Leslie, Alan, 69
Lewandowski, Joseph, 129
lifeworld. *See* umwelt
Lorenz, Konrad, 11, 37
Lubart, Todd, 128
Lurz, Robert, 78

MacLean, Evan, 83
MacNeil, David, 153–54
Mahurin, Ellen, 156
making-with. *See* sympoiesis
Marr, David, 34
Martin, Alia, 86–87, 93, 95
Maynard-Smith, John, 118
Mead, George Herbert, 99, 117–18
Meltzoff, Andrew, 45, 69, 71–72, 74
mental institutions (Gallagher concept), 133–34
mentalistic supposition, 72
mentalizing, 59–60, 62
metacognition, 72–73 79, 88, 118
methodological solipsism, 28
Miklósi, Ádám, 16, 83
mind, modular view of, 74–75
mindreading, 40, 46, 67; behavior reading, 69–70, 75, 78, 81–82, 84, 88–89, 111; continuum theory of, 6–7, 68–74, 76–77, 117; language and, 78–79, 89; mindblind, 91–93, 94t; nonhuman, 77–80, 82, 85, 90, 93, 114, 120
mirror neuron system, 71, 74
Mitchell, Robert, 7, 83, 99, 108–11, 112–15, 117, 119–20, 122, 127–30, 137, 141
monkeys. *See* primates, nonhuman
Moore, Derek, 75–76
Moore, Keith, 45, 71–72
Morgan, Conwy Lloyd, 24–25
Myin, Erik, 48

Nagel, Thomas, 27, 29–30, 121
natural environment, debates regarding, 11–12,

14–15, 19–20, 23, 37; unnatural-versus-natural model, 21
Nichols, Shaun, 69
Noë, Alva, 50–51, 135, 151

objective phenomenology, 29
octopi, 30–31, 105, 107
olfactory sense of dogs, 14, 42, 49–50
Oliver (dog of Laurel Braitman), 62
Other Minds (Godfrey-Smith), 30–31
oxytocin, 41–42

participatory sense-making (PSM), 57
Pavlov, Ivan, 37
Pellegrini, Anthony, 108
Penn, Derek, 79–80, 82, 84–85, 87–89, 91
Pfungst, Oskar, 16, 18
Philosophical Investigations (Wittgenstein), 120
phylogeny, 12
physical stance, 34
physical symbol systems (PSS), 43
Piaget, Jean, 113
Pilley, John, 39, 52
Plato, 100–101, 102; *Laws*, 100; *Republic*, 100
play, 7, 96, 98–100, 115; Burghardt's criteria of, 105; creativity and, 124, 126–28, 133–34, 141, 144; human, 101–4, 113, 165n1; improvisation and, 129, 150; interspecies, 109–12, 113, 127; nonhuman animal, 104–9; projects, 109–10, 113, 124, 128, 145; social, 112–13, 125; therapy, 102; work and, 100–101
pointing gesture, dog following of human, 38, 40, 54, 83
Politics (Aristotle), 100–101
Povinelli, Daniel, 79–80, 82, 84–85, 87–89, 91
predictive coding (PC), 139
predictive engagement (PE), 9, 49–50, 126, 139–43, 147
predictive processing (PP), 8, 126, 134–39, 143, 148, 150–51
Premack, David, 77–78
presabsence, 136
primates, nonhuman, 10, 71, 86, 95; chimps, 14, 39, 78, 80–81
problem of other minds, 27–29

Question of Animal Awareness, The (Griffin), 11

Racca, Anaïs, 41
radical philosophy of cognitive science, 3, 5, 36, 43, 47, 119, 135, 143 (*see also* enactivist approach to cognition); 4E approach (embedded, embodied, enacted, extended), 36
Ratcliffe, Matthew, 48
Reid, Alliston, 39
reptiles, 104
Republic (Plato), 100
Rescue (puppy saved by helicopter pilot), 1–2
Rivas, Jesús, 26
Rizzolatti, Giacomo, 71
Role of Play in Human Development, The (Pellegrini), 108–9
Rosch, Eleanor, 47
Rosenthal, David, 79
Ryall, Emily, 101
Ryle, Gilbert, 114

Santos, Laurie, 86–87, 93, 95
scent, 14, 42, 49–50
Scout (Super Bowl advertisement dog), 153–54
scribbling and bibbling, 146, 151, 156
Sebeok, Thomas, 14
second order social cognition (SSC), 93–95
Shipman, Pat, 55
signaling, 99, 117–20
Simmel, Marianne, 75
Simpson, M. J. A., 109
Simulation Theory, 6, 67, 69, 71–73, 97
siphonophore, 8, 145–48
Skinner, B. F., 11
Slaby, Jan, 49
Smith, Don, 1–2
sneak test, 54
social cognition, 35, 40, 65, 83–84, 93–94, 98, 127, 142. *See also* mindreading
Soproni, Krisztina, 83
Spaulding, Shannon, 72
sport-as-play, 101; agility sport, 58–59, 98, 118–19
SSM (strong sensorimotor model), 50
Staying with the Trouble (Haraway), 122–24
Sternberg, Robert, 128
Stich, Stephen, 69
still face experiments, 45
street dogs. *See* dogs, street
strong sensorimotor model (SSM), 50–51
subjective cognition, 63, 87–88, 99

suicidal non-human animals, 15, 62, 65
Supersizing the Mind (Clark), 138
Surfing Uncertainty (Clark), 135–38
Suter, Steve, 154
sympoiesis, 7, 9, 99, 123–24, 129–34, 148, 151–52, 155; enactivism and, 125–26; Joe-with-Mary, 144–45; siphonophores and, 8, 145–48

Takotsubo cardiomyopathy, 64–65
Target, Mary, 48
Tesla (dog of Michele Merritt), 106, 109–10
Theis, Kevin, 8
Theory of Mind (TOM). *See* mindreading
Theory Theory, 6, 67, 69–71, 72–73, 97
thick description, 114–16
"Thinking-Is-Moving" (Michele Merritt), 130–31
Thompson, Evan, 47–48, 109–10, 113–15, 117, 119–20, 122, 127, 141; *The Embodied Mind* (with Varela and Rosch), 47
Timberlake, William, 26
Tinbergen, Nikolaas, 11–12, 14, 25, 31, 37
Tomasello, Michael, 103
Trevarthen, Colwyn, 67, 103–4
Tronick, Edward, 45

Udell, Monique, 83, 85–86, 90, 93
umwelt: of dogs, 18, 31, 33, 37, 58, 93; theoretical concept, 4, 11, 13–14, 105, 140, 156

Varela, Francisco, 47
vector transformation, 60
video, of animals, 1–2
visual gestalt, 92
von Frisch, Karl, 11
von Osten, Wilhelm, 16–17
von Uexküll, Jakob, 14
Vonk, Jennifer, 80, 84
Vygotsky, Lev Semenovich, 101, 103

Westra, Evan, 92–93
When Species Meet (Haraway), 118–19
Wittgenstein, Ludwig, 100–101; *Philosophical Investigations*, 120
Wobber, Victoria, 83
wolves, 55, 62, 85
Woodruff, Guy, 77–78
Wynne, Clive, 25, 83

Zahavi, Dan, 29
Zaine, Isabela, 83

ANIMAL VOICES, ANIMAL WORLDS

ERIN MCKENNA, *Livestock:*
Food, Fiber, and Friends

ARNOLD ARLUKE AND ANDREW ROWAN, *Underdogs:*
Pets, People, and Poverty

ANNE BENVENUTI, *Kindred Spirits:*
One Animal Family

MICHELE MERRITT, *Minding Dogs:*
Humans, Canine Companions,
and a New Philosophy of Cognitive Science

CPSIA information can be obtained
at www.ICGtesting.com
Printed in the USA
LVHW090119090421
683868LV00005B/1125